SOURCES OF PLANET ENERGY, ENVIRONMENTAL & DISASTER SCIENCE

IMPACT OF GLACIER MELTING & CLIMATE CHANGE

National Seminar on SPEEDS-2022-23
held on
26th March, 2023

Bharat Raj Singh
Pramod Kumar Singh
Editors

Sujata Sinha
Ved Kumar
Co-Editors

I0426658

ISBN : 978-1-304-76342-6

Organized by

The Institution of Engineers (India)
U.P. State Centre, Lucknow
In
association with
School of Management Sciences, Lucknow
&
MNNIT Alumni Association Lucknow Chapter

The Institution of Engineers (India) School of Management Sciences,   MNNIT Alumni Association
U.P. State Centre, Lucknow   Lucknow   Lucknow Chapter

SOURCES OF PLANET ENERGY, ENVIRONMENTAL & DISASTER SCIENCE: IMPACT OF GLACIER MELTING & CLIMATE CHANGE

Published: Dec 2023
Edition: First
ISBN: 978-1-304-76342-6

Description:
The imperative to shift to sustainable alternatives is clear as widespread energy use contributes to global warming and glacier melting. Urgency arises in adopting eco-friendly solutions due to the significant harm caused by fossil fuels—coal, oil, and natural gas. These fuels lead to pollution, health issues, wildlife loss, and global warming emissions. Wind power offers clean electricity generation without toxic pollution or emissions. Despite being abundant and affordable, challenges like land use and impacts on wildlife persist. Solar and geothermal energy also play significant roles, with varying environmental impacts based on technology. Navigating these complexities emphasizes the need for a clear transition to renewable energy and sustainable solutions.

Editors:
  Bharat Raj Singh and **Pramod Kumar Singh**
Co-Editors:
  Sujata Sinha and **Ved Kumar**
CV Raman Center for Research & Innovation, under the aegis of **School of Management Sciences,**
  (Affiliated to Abdul Kalam Technical University)
  19th Km, Lucknow-Sultanpur Road, Kasimpur Biruha, Lucknow - 226501, India.
Website: www.smslucknow.ac.in **E-mail:** brsingh@smslucknow.ac.in

Publisher:
lulu Lulu Press Inc.
627, Davis Drive, Suite 300, Morrisville, NC 27560, USA
www.Lulu.com; Copyright © 2023 Lulu.com

PREFACE

It is widely recognized that our overall energy resources significantly impact the environment. Fossil fuels, including coal, oil, and natural gas, inflict considerable harm compared to renewable energy sources across various metrics. These traditional energy sources contribute to air and water pollution, negatively affecting public health, and leading to the loss of wildlife and habitats. They also contribute to issues such as water and land overuse and are major contributors to global warming.

The specific environmental impacts vary based on factors such as the technology employed, geographic location, and others. Harnessing wind power stands out as one of the cleanest and most sustainable methods of generating electricity. Wind energy production involves no toxic fumes or global warming emissions. Additionally, wind is abundant, inexhaustible, and cost-effective, making it a viable large-scale alternative to fossil fuels. However, it is not without its drawbacks, including challenges related to land use and impacts on wildlife and habitats.

Similarly, solar power from the sun presents a significant resource for generating clean and sustainable electricity. Despite its positive attributes, solar energy is not without environmental impacts, such as land use and habitat loss, water consumption, and the use of hazardous materials in manufacturing. The extent of these impacts varies depending on the scale of the system and the specific technology employed, whether it be photovoltaic (PV) solar cells or concentrating solar thermal plants (CSP).

Furthermore, geothermal power plants, also known as hydrothermal plants, are strategically located near geologic "hot spots" where molten rock is close to the Earth's crust, producing hot air. These plants vary in terms of the technology used for resource-to-electricity conversion (direct steam, flash, or binary) and the cooling technology employed (water-cooled or air-cooled). The environmental impacts associated with geothermal power generation are contingent on the specific conversion and cooling technologies utilized.

Other energy sources such as: biomass power plants exhibit similarities to fossil fuel power plants as both involve the combustion of a feedstock to generate electricity. Hydroelectric power encompasses both massive hydroelectric dams and small run-of-the-river plants. While large-scale hydroelectric dams are still being constructed in various parts of the world, it is unlikely that new facilities will need to be added to the existing ones in the future. Hydrokinetic energy, which includes wave and tidal power, comprises a range of energy technologies, many of which are either in the experimental stages or in early development.

Considering the aforementioned green energy resources and their widespread utilization through technological advancements, there is potential to significantly restore environmental damage, thereby reducing the substantial carbon content produced by fossil fuels today. However, addressing the complex challenge of global climate change requires unprecedented collaboration among entire industries, governments, citizens, communities, and more, both to mitigate and adapt to its effects.

A one-day National Seminar titled "Sources of Planet Energy, Environment and Disaster Science: Impact of Glacier Melting & Climate Change (SPPEDS-2022-23)" took place on March 26, 2023, at The Institute of Engineers, UP State Centre, Lucknow, in association with SMS Lucknow and MNNIT Alumni Association, Lucknow Chapter. The seminar aimed to explore climate risks arising from Glacier Melting in the 21st century and potential remedial actions.

To facilitate in-depth discussions, technical sessions were organized, featuring contributions from intellectual leaders in the fields of science, technology, academia, and industry. These sessions focused on vibrant issues and solutions related to the following themes:

1. Glacier Melting on Arctic Sea
2. Glacier Melting on Antarctica Sea
3. Glacier Melting on Himalayas
4. Global Warming
5. Climate Change
6. Disaster Management
7. Waste Management
8. Renewable Energy
9. Others

In response to the call for papers, we received a total of 50 submissions spanning various themes, with 35 of them being accepted and 20 ultimately presented. We urged all presenters and keynote speakers to shed light on innovative ideas addressing the challenging issues surrounding global warming and climate damage.

This book comprises a curated selection of 15 research papers, encompassing four (4) from Themes 1 to 3 (Glacier Melting on Arctic Sea, Glacier Melting on Antarctica Sea, and Glacier Melting on Himalayas), three (3) from Themes 4 to 7 (Global Warming, Climate Change, Disaster Management, and Waste Management), two (2) from Theme 8 (Renewable Energy), and six (6) from Theme 9 (Other topics related to Climate and Disaster).

I extend my heartfelt gratitude to the editorial board, the authors of the research papers, and all stakeholders who provided unwavering support and guidance throughout the seminar. A special acknowledgment is reserved for Mr. Sharad Singh, Secretary and CEO of SMS Lucknow, whose motivation and mentorship played a pivotal role in steering the entire team.

At the end, words fail to express my reverence for the almighty, without whose divine intervention, this seminar programme would not have culminated in its final form.

Prof. (Dr.) Bharat Raj Singh,
Editor-in-Chief & Seminar Co-Chair
Director General (Technical), SMS, Lucknow

ACKNOWLEDGMENT

The success of an event is directly tied to the collaborative efforts of its team, coupled with meticulous planning and the execution of innovative ideas. The outstanding success of this seminar is a testament to the hard work of the team that brought together distinguished individuals from both research backgrounds and industry to the platform of SPEEDS.

We take great pride in acknowledging that the dedication of each and every individual involved in SPEED 2022-23 contributed significantly to its success. We are deeply grateful to the Institution of Engineers (India) U.P. State Center and MNNIT Alumni Association Lucknow Chapter for their continuous support and sponsorship of this seminar.

Our sincere appreciation goes to all the dignitaries and presenters who generously shared their time and insights on various topics and sub-topics of the seminar titled: "Sources of Planet Energy, Environmental and Disaster Science: Impact of Glacier Melting" (SPEEDS 2022-23) held on March 26, 2023.

Special thanks to Mr. R. K. Singh, a member of the UP Pollution Control Board for gracing the event as the Chief Guest during the inaugural session. We extend our gratitude to Prof. (Dr.) Jyotsna Singh, Director of the Centre for Excellence in Renewable Energy, Lucknow University, for being a Guest of Honour and keynote speaker, as well as Er. B. C. Roy, President of MNNIT Alumni Association, Lucknow Chapter, for gracing the valedictory session as Chief Guest.

Our heartfelt thanks also go to Prof. Ashok Kumar Tiwari, General Secretary of MNNIT Alumni Association, Lucknow Chapter, who served as the Guest of Honour for the valedictory session, making a valuable contribution to the seminar. We express our deepest regards to Er. R. K. Trivedi, Immediate Past Chairman of the Institution of Engineers (India): UP State Centre Lucknow, Keynote speaker Dr. P. K. Bharati, HOD- ME, Integral University Lucknow, and Dr. Niraj Gupta, Prof. & Dean, SRM University, Lucknow.

We are equally grateful to Dr. Venkatesh Dutta, Assistant Professor at SES, BBAU, Lucknow, and Dr. Suresh Chandra Bajpai, Director of Futuris Energy Pvt. Ltd., for their insightful contributions during the seminar. Special mention is due to our patrons, Dr. M.P. Singh, Executive Secretary SMS Society, Mr. Sharad Singh, Secretary & CEO, SMS Lucknow, Prof. (Dr.) Manoj Mehrotra, Director SMS Lucknow, and Prof. (Dr.) Dharmendra Singh, Associate Director SMS, Lucknow and Seminar Coordinator, for their unwavering support and guidance.

We extend our deepest gratitude to Prof. (Dr.) P.K. Singh, Seminar Co-Convener, Dr. Ved Kumar, Seminar Co-Coordinator, Dr. Jagdish Singh, Chief General Manager, Mr. Surendra Srivastava, General Manager (Corporate Relations) at SMS, Lucknow, for their exceptional support and cooperation.

Our sincere thanks to the Advisory Committee Members, Executive Committee Members, and all Faculty Members associated with us directly or indirectly. Special mention goes to Mrs. Sujata Sinha, Dr. Shrinkhla Srivastava, and Mr. Amit Kumar Srivastava (DTP) for their cooperation and assistance since the inception of this seminar.

We express our deepest regard to our leading sponsors, the Institution of Engineers and Ultra Tech Cement, without whom this program would not have achieved such success. Last but not least, our appreciation goes to all the Delegates, Participants, Paper Presenters, and Media personnel for their invaluable support, which played a crucial role in elevating this seminar to its heights.

Prof. (Dr.) Bharat Raj Singh,
Editor-in-Chief & Seminar Co-Chair,
Director General (Tech.),
SMS, Lucknow

One Day National Seminar on

Sources of Planet Energy, Environmental & Disaster Science
(SPEEDS-2022-23):
Impact of Glacier Melting & Climate Change
held on
March 26th, 2023

Report and Recommendations on the Proceedings of the National Seminar

1. Deliberations during the Morning Session on March 26, 2023

1.1 Professor (Dr.) Manoj Mehrotra, Director-SMS Lucknow, extended a warm welcome to the Chief Guest Mr. R. K. Singh, Guest of Honor Dr. Jyotsna Singh, Keynote Speaker Dr. P. K. Bharti, and other esteemed dignitaries, delegates, and participants. In his address, he highlighted that the Sustainable Development Goals (SDGs), initiated in 2015 in the United States, particularly SDGs 13, 14, and 15, focus on addressing global warming and environmental damage. Emphasizing the need to adhere strictly to these guidelines, he underscored the importance of safeguarding against climate damage caused by the impact of global warming and glacier melting.

1.2 Er. R.K. Trivedi, Immediate Past Chairman, assumed the chair for the Inaugural session, extending a warm welcome to the guests and audience. During his address, he commended the decisions made during the 2015 Paris Summit Agreement, aiming to limit the temperature rise to not exceed 1.50 degrees Celsius, as discussed by the Union Secretary. He emphasized the potential dangers to the world if this limit were surpassed.

1.3 Professor (Dr.) Bharat Raj Singh, Director General (Tech.) and Seminar Co-Chair and Convener of SPEEDS 2022-23, elucidated on the theme of the One Day National Seminar: "Sources of Planet Energy, Environmental & Disaster Science: Impact of Glacier Melting & Climate Change." He highlighted the accelerated melting of glaciers due to manmade global warming, resulting in severe climate damage and rising sea levels. Addressing coastal submergence, cyclones, tsunamis, and snowfall, he advocated for controlling these impacts through the utilization of renewable energy sources and tree planting. He proposed the adoption of the slogan 'Save Earth, Save Life.'

1.4 Dr. P.K. Bharti, the Keynote Speaker, presented a case study on a 1-megawatt solar power plant and the prospects of solar power. He emphasized that globally installed such plants could supply sufficient electrical energy, meeting the demand without relying on hydrocarbons.

1.5 Dr. Jyotsana Singh, Guest of Honour and Director of the Centre of Excellence in Renewable Energy, Education & Research at Lucknow University, discussed the production and meaningful utilization of energy. Stressing the importance of protecting nature, she invoked Mahatma Gandhi's slogan, 'There is enough for everyone's need but not enough for everyone's greed.' She addressed ways to reduce carbon footprints and the impacts of climate change. (v)

1.6 Chief Guest Dr. R.K. Singh, Technical Advisor at the Uttar Pradesh Pollution Control Board, illustrated the evolution of Earth using the example of the movie Jurassic Park. Expressing concerns about climate change and its effects on humans, he discussed the vision of the central government, policies of Niti Aayog, and the banning of plastic use in the state, which he noted was not strictly followed. Dr. Singh also discussed the flow of the Gomti River and advised constructing a barrage near Gaughat for its free flow.

1.7 Jaswant Singh, Honorary Secretary, IEI, UP State Centre, extended a vote of thanks to the Chief Guest and other participants. He suggested incorporating the recommendation of building a barrage near Gaughat and acknowledged Professor Jyotsna Singh's concerns about climate change and Dr. P.K. Bharti's advice on including a solar plant in the recommendations.

2.0 Deliberations in the Afternoon Session on March 26, 2023

Technical Session I, chaired by Dr. P. K. Bharti, Keynote Speaker (HOD-Mechanical Integral University, Lucknow)

2.1 Dr. Priyank Sharma and Dr. Kailash Pati presented on various types of environmentally friendly CSR activities. They emphasized that companies strive to create goodwill in the eyes of society and target customers through corporate social responsibility. Sustainable organizations must maintain and preserve the natural environment in which they operate.

2.2 Dr. Bharat Raj Singh, Director General (Tech), presented the paper titled 'Melting of Himalayan Glacier Causing Climate Damage and Impacting India with Heavy Snowfalls, Cold Waves, Cyclones, and Cracks in Hills.' He cautioned about the permafrost effect leading to landslides and loss of life near the Himalayan glacier region due to snowfalls.

2.3 Mr. Sanjeev Kumar Pandey from the Department of Civil Engineering presented the paper 'Analysis of Physico-Chemical Characteristics of Water Quality Discharged from CETP, Banthra, Unnao.' The discussion focused on the discharge of chemicals into the Ganga River, causing pollution.

2.4 Mr. Himanshu Minotra from the Department of Mechanical Engineering presented the paper 'Design of a Movable Solar Panel on a Waste Bin.' He discussed the complete design of a solar panel on a waste bin, including its base stand, battery stand, and wheels, created using SOLID WORKS software.

3.0 Invited Talks

3.1 In the post-lunch session, Dr. Venkatesha Dutta from the School of Earth & Environmental Sciences, BBAU, Lucknow, presented the paper 'Restoring Smaller Rivers in Uttar Pradesh.' He discussed the correlation between people, climate, and habitat, addressing ecosystem dehydration and economic impacts. Ms. Shipra Pathak, known as the Water Woman of India, was invited to share her insights on the spiritual connection and importance of the Gomti River. Mementos were presented to both speakers.

3.2 In the second invited talk, Dr. Suresh Chandra Vajpayee discussed various aspects of renewable energy use.

Technical Session II, chaired by Dr. Suresh C. Bajpayi, Invited Speaker

4.0 Technical Session-II

4.1 Mr. Ashok Sengupta from the Department of Management, School of Management of Science Lucknow, presented the paper 'Transforming Land Usage by Agricultural Engineering – A Study of Saudi Arabia.' He highlighted Saudi Arabia's agricultural reforms, transforming it from a food deficit country to one of the largest wheat producers.

4.2 Mr. Prasoon Mishra & Mr. Himanshu Baranwal, students of Computer Science and Engineering, presented the paper 'E-commerce Web Application Using Mern Technology,' detailing an E-Commerce Shopping Cart application.

4.3 Ms. Samreen Siddiqui, a student from Amity School of Applied Science, Amity University, presented the paper 'Quantum Dots for Solar Cell,' focusing on the use of quantum dots as photovoltaic absorbing material in solar cells.

4.4 Dr. Ved Kumar from the Department of Humanity and Applied Sciences, School of Management Sciences, Lucknow, presented the paper 'Green Synthesis of Titania and Titania-Silver Nanoparticle from Novel Plant Extract of Origanum Majorana.' He discussed the eco-friendly synthesis of TiO Nanomaterials.

5.0 Valedictory Session

5.1 Er. B.C. Roy, President of MNNIT Alumni Association, Lucknow Chapter, thanked everyone and highlighted the work of the Alumni Association and the Lucknow chapter.

5.2 Dr. Bharat Raj Singh, the Director General (Technical) and convener of the event, delivered concluding remarks. He expressed gratitude to all 20 presenters who shared their innovative perspectives on their chosen topics. Special appreciation was extended to the technical session chairs, Dr. P.K. Bharti and Dr. Suresh Chandra Bajpayai, for their smooth facilitation of the sessions.

The house was also deliberated on recommendations and sought the participants' consent to forward them to the government and various appropriate authorities. This collective effort aimed at encouraging adoption for the benefit of society. Appreciation was extended to the Institute of Engineers UP State Centre Lucknow and the MNNIT Alumni Association, Lucknow Chapter, for their invaluable support.

5.3 Dr. Usha Vajpayee, Chief Guest of the Valedictory Session, spoke on the topic of Solid waste, Kitchen waste, and Harnessing energy from nature for life systems.

5.4 Associate Director Dr. Dharmendra Singh proposed the Vote of Thanks. Mementos were presented to all presenters, the Guest of Honour, and the Chief Guests of Inaugural and Valedictory Sessions.

Recommendations

1. It is recommended that addressing climate change is a straightforward task, emphasizing a shift in our lifestyle, aligning with the mission initiated by our Hon'ble Prime Minister. Therefore, establishing a center in Lucknow, Uttar Pradesh, dedicated to advocating and advancing the mission charter within society, is essential.

2. Encouraging the installation of Small Solar Power Plants, generating up to 1 Megawatt, in major cities is crucial to meet the growing demand for additional electric power.

3. Development of River Fronts in metropolitan areas, coupled with extensive plantation along riverbeds, is proposed to mitigate air pollution and promote tourism.

4. The construction of a barrage at Gaughat, Lucknow, is recommended to ensure the cleanliness and greenery of the upstream area of the Gomti River within the city, covering approximately eight kilometers.

5. Waste management initiatives within urban centers should be prioritized to address the challenges posed by increasing waste.

6. Agricultural reforms are suggested, especially in areas where land is diminishing due to residential construction. Encouraging hydroponic farming can be a sustainable solution in such regions.

National Seminar on
"Sources of Planet Energy, Environmental & Disaster Science Impact of Glacier Melting & Climate Change (SPEEDS-2022-23) held on 26ᵗʰ March, 2023

Chief Patron
Prof. (Dr.) Alok Kumar Rai,
Hon'ble Vice Chancellor
Dr. A.P.J. Abdul Kalam Technical University,
U.P. Lucknow

Patron
Dr. M.P. Singh,
Executive Secretary
School of Management Sciences,
Varanasi

Patron
Shri Sharad Singh,
Secretary & CEO,
School of Management Sciences,
Lucknow

Seminar Chair/Co-Chair
Prof. (Dr.) Manoj Mehrotra
Director, SMS, Lucknow
Prof. (Dr.) Bharat Raj Singh
DG(Tech.) SMS, Lucknow

Seminar Coordinator
Prof. (Dr.) Dharmendra Singh
Associate Director, SMS, Lko.

Convener
Prof. (Dr.) Pramod Kumar Singh
Dean Student Welfare, SMS, Lko.

Members
Mr. Surendra Srivastava, G.M. (Corp. Rel.), SMS, Lucknow

Dr. Hemant Kumar Singh, Dean Engg. SMS, Lucknow

Prof. (Dr.) Amarjeet Singh, HOD-EE, SMS, Lucknow

Dr. Ajay Singh Yadav, HOD-HAS, SMS, Lucknow

Dr. Asha Kulshreshtha, HOD-CE, SMS, Lucknow

Dr. Shrinkhala Srivastava, Coordinator-HAS, SMS, Lucknow

Mr. Amod Pandey, Principal-Diploma, SMS Lucknow

Mr. Pankaj Yadav, HOD-ME, SMS Lucknow

Dr. Anod Kumar Singh, Coordinator-B.Sc, SMS Lucknow

Mr. Anoop Kumar Singh, Assistant Prof., SMS Lucknow

Organized by

The Institution of Engineers (India) U.P. State Centre, Lucknow
In
association with
School of Management Sciences, Lucknow
&
MNNIT Alumni Association Lucknow Chapter

The Institution of Engineers (India)
U.P. State Centre, Lucknow

School of Management Sciences,
Lucknow

MNNIT Alumni Association
Lucknow Chapter

(ix)

NationalAdvisory Committee

Dr. Karl Cheng, Chairman, Innotest Inc., Science Park, Hsinchu, Taiwan

Prof. (Dr.) Adnan Middilli, Vice Rector for Research and Innovation, Recep Tayyrip Erdogan University, Turkey

Prof. (Dr.) Onkar Singh, V.C. Veer Madho Singh Bhandari Uttarakhand Technical University, Dehradun (Uttarakhand)

Prof. N.B. Singh, V.C. Khwaja Moinuddin Chishti Language University, Lucknow

Prof. (Dr.) D.S. Chauhan, V.C. GLA, Mathura

Prof. (Dr.) Kripa Shanker, Ex-VC, Dr. A.P.J. Abdul Kalam Technical University, Lucknow

Shri Alok Kumar, Director General, Council of Science & Technology, U.P.

Dr. Vineet Kansal, Director, IET, Lucknow.

Prof. (Dr.) S.P. Shukla, Director, Rajkiya Engineering College, Banda.

Prof. (Dr.) P.N. Jha, Director, SMS, Varanasi.

Prof. (Dr.) D. Buddhi, Director, SIET, Dehradun.

Prof. (Dr.) K. Hansraj, Department of ME, DEI, Agra.

Prof. (Dr.) Jagbir Singh, Director (Architecture) Amity University, Lucknow.

Organizing Committee

Er. V.B. Singh, National Vice President, IE(I) Kolkatta

Er. Masarrat Noor Khan, Chairman IE(I), UP State Centre

Dr. Jaswant Singh, Hon. Secretary IE(I), UP State Centre

Er. S.C. Mehra, FIE, SCC, IE(I), UP State Centre

Er. Vijay Pratap Singh, FIE, SCC, IE(I), UP State Centre

Er. K.P. Tripathi, FIE, SCC, IE(I), UP State Centre

Er. Satya Prakash, FIE, SCC, IE(I), UP State Centre

Er. B.C. Rai, President, MAA, Lucknow Chapter

Prof. Ashok Kumar Tiwari, General Secretary, MAA, Lucknow

Er. Manjeet Singh, Senior Vice President, MAA, Lucknow

Dr. Jagdish Singh, Director (Admin & Admissions), SMS, Lucknow

Proceedings International Seminar on "Sources of Planet Energy, Environmental & Disaster Science: Impact of Glacier Melting & Climate Change (SPEEDS-2022-23)

CONTENTS

RENEWABLE ENERGY

OTHERS (RELATED TO CLIMATE AND DISASTER)

Glimpses of Memorial Event - SPEEDS 2022-23

Lamp Lighting Ceremony

Chief Guest Dr. R. K. Singh being offered the
Bouquet by Director SMS
Prof. (Dr.) Manoj Mehrotra

3. Dr. P. K. Bharati being presented with a
Bouquet by DG(Tech.) Dr. Bharat Raj Singh

Er. R. K. Trivedi being presented with a
Bouquet by DG(Tech.) Dr. Bharat Raj Singh

Inaugural Session

Glimpses of Memorial Event - SPEEDS 2022-23

DG(Tech.) Dr. Bharat Raj Singh Delivering
the Convenor's note

Unveiling of Samriddhi

Key Note Speaker Dr. Venketesh Dutta
delivering his Keynote Address

Water Woman of India at SPEEDS 2023

Dr. Priyank Sharma making his
presentation

Glimpses of Memorial Event - SPEEDS 2022-23

Session Chair, Dr. Suresh Chandra Bajpayai being offered the Bouquet by Director SMS Prof.(Dr.) Manoj Mehrotra

Mr. Ashok Sen Gupta Making his presentation

Er. B.C. Roy President, MNNIT Alumni Association, Lucknow Bouquet by DG(Tech.) Dr. Bharat Raj Singh

Dr. Jaswant Singh being presented with a momento

Dr. Ved Kumar being presented with a momento for his presentation

The Rapid Melting of Arctic Ice Become a Major Cause of Cyclonic Snowfall in USA

Bharat Raj Singh

School of Management Sciences, Lucknow (affiliated with AKTU),

e-mail: brsinghlko@yahoo.com

ABSTRACT

The massive US winter storm has left millions without power since last Friday, December 22, 2022, and canceled all their vacation plans. Power outages darkened more than 1.4 million homes and businesses, while thousands of US flights were cancelled. This fierce winter storm, called Elliot by forecasters, intensified into a bomb cyclone near the Great Lakes on Friday, December 22, 2022, bringing high winds and blizzards from the northern plains to western and upstate of New York. It has become serious cause of killing in the form of floods, flash-freezing and travel intrusions it took a terrible toll. This led to the cancellation of over 5,700 flights by airlines, leaving thousands of passengers stranded at airports. Icy weather or accidents disrupted travel on roads, and officials in parts of Indiana, Michigan, New York and Ohio urged motorists to avoid unnecessary travel. Transportation Secretary informed that the US aviation system is "operating under enormous pressure". On December 21, 2022 alone, about 10% of US flights were canceled. The devastating effect of the storm spread over a width of 3,200 km, which means heavy snowfall from Texas to Maine and a drop in temperature from (-) 45 to (-60) 60 degree centigrade, making life miserable for the people. Authorities ordered cars off the roads as US forecasters warned of "extreme impacts" from the potentially cyclonic blizzard in central and eastern parts of the country. This paper evaluates the reasons for such devastating climatic conditions and what kind precautions people has to take in future.

Keywords: *Melting of Arctic, winter storm, terrible toll, flash-freezing, snowfall.*

1. INTRODUCTION

The global climate has undergone significant changes, leading to unprecedented shifts in weather patterns. One of the alarming consequences of climate change is the rapid melting of Arctic ice, which has emerged as a major contributor to cyclonic snowfall events in the United States. This phenomenon poses a serious threat to communities, infrastructure, and livelihoods, necessitating a comprehensive understanding and proactive management strategies.

In the annals of Earth's climatic history, recent years stand out as an era marked by profound and unsettling transformations. Global climate change, propelled by anthropogenic activities, has ushered in an era of unprecedented shifts in weather patterns, challenging the resilience of ecosystems and human societies alike. Within this complex mosaic of climatic changes, a particularly alarming consequence has surfaced—the rapid melting of Arctic ice. This phenomenon, once relegated to scientific discussions, has now become a tangible and pressing threat, with far-reaching implications, especially concerning cyclonic snowfall events in the United States.

The Arctic, a vast and remote expanse at the Earth's northernmost reaches, has long been a sentinel of climate change. The region's ice cover, a crucial component of the planet's climate system, acts as both a barometer and a driver of global weather patterns. However, the relentless warming of the Earth's atmosphere due to the accumulation of greenhouse gases has precipitated a swift and dramatic response in the Arctic as shown in Figure 1. The consequences of this response reverberate far beyond the polar circle, impacting weather systems thousands of miles away.

Figure 1: Fast melting of Arctic Sea

The focus of this discourse is the intricate interplay between the rapid melting of Arctic ice and the surge in cyclonic snowfall events experienced in the United States. While the contiguous U.S. may seem geographically distant from the Arctic, the intricate web of atmospheric and oceanic connections links the two regions in ways that have profound implications for the daily lives of Americans. Cyclonic snowfall, characterized by intense storms combining snow and wind, has become a recurrent and formidable force, disrupting routines, straining infrastructure, and challenging the very fabric of communities [1].

As the impacts of cyclonic snowfall events become more pronounced, understanding the dynamics of this climatic shift is imperative as shown in Figure 2. The stakes are high, encompassing not only the physical safety of individuals but also the resilience of critical infrastructure and the sustainability of livelihoods. Thus, the imperative for a comprehensive understanding of the interplay between Arctic ice melt and cyclonic snowfall events is clear, as is the need for proactive management strategies to mitigate the impending threats.

The complexity of this issue demands a multi-faceted exploration, considering not only the scientific intricacies of climate interactions but also the socio-economic dimensions of the challenges posed. As we delve into the heart of this matter, we embark on a journey to unravel the threads connecting the Arctic's icy expanses to the cyclonic snowfall events wreaking havoc in the United States. Through this exploration, we seek not only to comprehend the mechanisms driving these changes but also to elucidate strategies that can safeguard communities, fortify infrastructure, and preserve livelihoods in the face of an evolving climate [2].

Figure 2: Heavy Snow fall in USA

In the subsequent sections, we will delve into the status of the increase in snowfalls and cyclonic rains, examining the tangible manifestations of these climatic shifts. Following this, we will explore various methods designed to manage and protect livelihoods, recognizing the need for a holistic approach that combines technological innovation, community engagement, and sustainable practices. Finally, drawing upon the insights gleaned, we will arrive at comprehensive conclusions that underscore the urgency of collective action and underscore the significance of these findings in the broader context of climate change adaptation and mitigation.

The narrative that unfolds is not merely a scientific investigation but a call to action—a recognition that the convergence of climatic forces demands a concerted effort to secure a sustainable and resilient future. As we navigate the intricacies of Arctic ice melt and cyclonic snowfall, we are compelled to confront the pressing reality that the choices we make today will shape the climate challenges of tomorrow.

2. STATUS OF INCREASE IN SNOWFALLS AND CYCLONIC RAINS

The Earth's climate is undergoing unprecedented changes, with observable shifts in weather patterns that have far-reaching consequences. A significant aspect of these changes is the rapid melting of Arctic ice, a consequence of human-induced climate change. This phenomenon has emerged as a major contributing factor to the intensification of snowfall and cyclonic rains in various regions across the globe, presenting substantial challenges, particularly in the United States.

2.1 Arctic Ice Melting and Climate Disruption

The Arctic, long considered a bellwether for global climate changes, is experiencing accelerated ice melt due to rising temperatures. This melting has disrupted traditional climate systems, triggering a cascade of effects that extend beyond the polar region. The interconnected nature of the Earth's climate means that changes in the Arctic have repercussions on weather patterns worldwide [3].

2.2 Intensification of Snowfall and Cyclonic Rains

The consequences of Arctic ice melt manifest in the intensification of snowfall and cyclonic rains, particularly in regions like the United States. The increased frequency and severity of these weather events pose significant challenges to communities, infrastructure, and public safety [4-5].

2.3 Challenges and Impacts

The consequences of heightened snowfall and cyclonic rains are multifaceted. Infrastructure damage is a prevalent issue, with extreme weather events straining buildings, roads, and utility systems. Disruptions to transportation networks, including delays and closures, further compound the challenges faced by communities. Moreover, the heightened risk to public safety, including the potential for accidents and injuries, underscores the urgency of addressing the impacts of these climatic shifts [6-7].

2.4 Crucial Need for Understanding

Understanding the status of these changes is paramount for developing effective strategies to mitigate their impact and protect communities. Robust scientific research, coupled with real-time monitoring and data analysis, is essential to grasp the intricacies of the evolving climate dynamics. This understanding forms the foundation for developing adaptive strategies that can enhance resilience and minimize the vulnerabilities associated with increased snowfall and cyclonic rains [8].

Thus, the status of the increase in snowfalls and cyclonic rains, fueled by the rapid melting of Arctic ice, demands a comprehensive and interdisciplinary approach. The amalgamation of scientific research, policy initiatives, and community engagement is essential to effectively address the challenges posed by these climatic shifts. As we navigate the complexities of a changing climate, it is imperative to draw on a wealth of knowledge to inform strategies that can safeguard communities and build a resilient future.

3. METHODS TO MANAGE AND PROTECT LIVELIHOODS

Addressing the challenges posed by cyclonic snowfall requires a multifaceted approach that combines mitigation and adaptation strategies. Some key methods to manage and protect livelihoods include:

3.1 Early Warning Systems

One crucial strategy to address the challenges posed by increased snowfall and cyclonic rains, resulting from the rapid melting of Arctic ice, is the implementation of advanced Early Warning Systems (EWS). These systems play a pivotal role in enhancing community preparedness by utilizing state-of-the-art meteorological technologies and communication systems to provide timely and accurate forecasts. The goal is to empower communities to take proactive measures in anticipation of impending weather events, thereby reducing the potential impact on infrastructure and ensuring public safety.

3.1.1 Meteorological Technologies: The advancement of meteorological technologies is essential for improving the precision and reliability of weather forecasting. Cutting-edge instruments, such as satellite-based observation systems and high-performance computer models, enable scientists to monitor atmospheric conditions with unprecedented detail. These technologies enhance the ability to detect and analyze the complex interactions that lead to increased snowfall and cyclonic rains [9-10].

3.1.2 Communication Systems: Effective communication is paramount in translating meteorological forecasts into actionable information for communities. Early Warning Systems leverage various communication channels, including broadcast media, social media, and mobile applications, to disseminate timely and accurate information. This ensures that residents receive warnings and instructions promptly, allowing them to make informed decisions about evacuation, sheltering, or other protective measures [11-12].

3.1.3 Community Preparedness: The ultimate aim of Early Warning Systems is to enhance community preparedness. This involves not only the provision of information but also fostering a culture of resilience within communities. Educational campaigns, drills, and community engagement initiatives are integral components of this preparedness effort. By actively involving residents in the process, Early Warning Systems empower them to understand the risks, respond effectively, and collaborate with local authorities in implementing protective measures [13-14].

Incorporating advanced meteorological technologies and communication systems into Early Warning Systems represents a proactive and adaptive approach to the challenges posed by changing climate dynamics. By fostering a culture of preparedness and leveraging scientific advancements, communities can navigate the complexities of increased snowfall and cyclonic rains with resilience and efficacy.

3.2 Infrastructure Resilience

A critical strategy to mitigate the impact of increased snowfall and cyclonic rains, driven by the rapid melting of Arctic ice, is to invest in infrastructure resilience. Resilient infrastructure is designed to withstand extreme weather conditions, providing a robust defense against the physical and economic consequences of severe climate events. This comprehensive approach involves reinforcing buildings, upgrading transportation networks, and improving drainage systems to enhance the overall resilience of communities.

3.2.1 Reinforcing Buildings: Strengthening the resilience of buildings is essential to ensure their ability to withstand the destructive forces of snowstorms and cyclones. This involves adopting and enforcing building codes that consider the specific risks associated with increased precipitation and extreme weather events. Additionally, innovative construction materials and techniques, informed by climate-resilient design principles, contribute to the structural integrity of buildings in the face of changing climatic conditions [15-16].

3.2.2 Upgrading Transportation Networks: Resilient transportation networks are pivotal for maintaining connectivity and accessibility during and after extreme weather events. Investments in upgrading roads, bridges, and public transit systems take into account the changing climate patterns, ensuring that these critical components of infrastructure can withstand the impacts of snowstorms and cyclones. This includes the implementation of adaptive measures such as elevated roadways, reinforced bridges, and improved drainage along transportation routes [17-18].

3.2.3 Improving Drainage Systems: Effective drainage systems are crucial for managing the increased precipitation associated with cyclonic rains. Investing in drainage infrastructure, including stormwater management systems and flood control measures, helps prevent inundation and reduces the risk of infrastructure damage. Sustainable drainage practices, such as green infrastructure and permeable surfaces, contribute to effective water management while promoting environmental sustainability [19-20].

Investing in infrastructure resilience is a forward-looking approach that recognizes the inevitability of climate change impacts. By reinforcing buildings, upgrading transportation networks, and improving drainage systems, communities can adapt to the challenges posed by increased snowfall and cyclonic rains, ensuring the long-term sustainability and functionality of critical infrastructure.

3.3 Community Education and Preparedness

In addressing the challenges posed by cyclonic snowfall, a fundamental strategy is the implementation of community education and preparedness initiatives. By conducting awareness campaigns, communities can be informed about the specific risks associated with cyclonic snowfall and equipped with essential information on evacuation routes, emergency shelters, and preparedness measures. This proactive approach is crucial for enhancing community resilience and ensuring the safety of residents during extreme weather events.

3.3.1 Awareness Campaigns: Conducting targeted awareness campaigns is essential to educate communities about the specific risks and challenges posed by cyclonic snowfall. These campaigns utilize various communication channels, including social media, community events, and educational materials, to disseminate information. The aim is to increase public understanding of the potential impacts of cyclonic snowfall and foster a culture of preparedness within the community [21-22].

3.3.2 Information on Evacuation Routes: Providing communities with clear and accessible information on evacuation routes is crucial for facilitating timely and orderly evacuations during cyclonic snowfall events. This involves developing comprehensive evacuation plans that are communicated to the public through various means, including signage, community meetings, and online platforms. Well-defined evacuation routes contribute to minimizing the risks to residents and ensuring an efficient response to evolving weather conditions [23-24].

3.3.3 Emergency Shelters: Establishing and communicating information about emergency shelters is a vital component of community preparedness. This involves identifying suitable shelter locations, ensuring they are well-equipped to handle the needs of residents, and disseminating this information widely. In addition, community engagement efforts can encourage residents to familiarize themselves with the locations of nearby shelters, fostering a sense of collective responsibility for community safety [25-26].

3.3.4 Preparedness Measures: Educating communities about specific preparedness measures is crucial for empowering residents to take proactive steps in safeguarding themselves and their property during cyclonic snowfall. This may include guidance on securing homes, preparing emergency kits, and understanding the importance of early evacuation. By fostering a sense of personal responsibility and readiness, communities can better withstand the impacts of cyclonic snowfall [27-28].

Therefore, community education and preparedness are integral components of a comprehensive strategy to mitigate the impact of cyclonic snowfall. By enhancing public awareness and providing essential information on evacuation routes, emergency shelters, and preparedness measures, communities can build resilience and respond effectively to the challenges posed by extreme weather events.

3.4 Natural Resource Management

As a critical facet of mitigating the impact of extreme weather events, including cyclonic snowfall driven by the rapid melting of Arctic ice, the implementation of sustainable natural resource management practices stands out as an essential strategy. This approach involves safeguarding vital ecosystems, such as forests and wetlands, which play a crucial role in enhancing resilience and mitigating the adverse effects of climatic shifts.

3.4.1 Forests as Climate Resilient Assets: Forests are pivotal in climate regulation, sequestering carbon, moderating temperatures, and influencing precipitation patterns. Sustainable forest management practices, encompassing afforestation, reforestation, and reduced deforestation, contribute to climate resilience. These measures not only enhance the adaptive capacity of ecosystems but also provide societal benefits, including improved water quality, biodiversity conservation, and carbon sequestration [29-30].

3.4.2 Wetlands for Flood Mitigation: Wetlands act as natural buffers against floods and extreme weather events by absorbing and slowing down water flow. Sustainable wetland management involves preserving existing wetlands, restoring degraded ones, and avoiding harmful practices such as draining or filling. These actions not only reduce the risk of flooding but also enhance water quality, support biodiversity, and provide valuable habitats for various species [31-32].

3.4.3 Biodiversity Conservation and Resilience: Biodiverse ecosystems, supported by effective natural resource management, exhibit greater resilience to extreme weather events. Preserving biodiversity contributes to ecosystem stability, ensuring that various species can adapt to changing conditions. Protected areas, sustainable land-use planning, and conservation efforts are integral to maintaining biodiversity and reinforcing the capacity of ecosystems to withstand climatic stressors [33-34].

3.4.4 Carbon Sequestration and Climate Mitigation: Natural resource management practices contribute significantly to carbon sequestration, a key strategy in mitigating climate change impacts. Forests, wetlands, and other ecosystems act as carbon sinks, capturing and storing atmospheric carbon dioxide. Sustainable practices, such as reforestation and afforestation, enhance the capacity of ecosystems to sequester carbon, thereby mitigating the overall impact of extreme weather events [35-36].

Thus, implementing sustainable natural resource management practices is integral to building resilience against the impacts of extreme weather events. By protecting and restoring ecosystems, communities can harness the inherent benefits of forests and wetlands, contributing to climate adaptation, biodiversity conservation, and overall environmental sustainability.

4. CONCLUSIONS

The rapid melting of Arctic ice has become a significant driver of cyclonic snowfall in the United States, posing considerable challenges to the well-being of communities and their livelihoods. Addressing this issue requires a collaborative effort involving governments, communities, and various stakeholders. By implementing early warning systems, enhancing infrastructure resilience, promoting community education, and adopting sustainable practices, it is possible to mitigate the impact of cyclonic snowfall and build a more resilient society. The factors to be addressed are given below:

a) Arctic Ice Melting as a Key Contributor: The acceleration of Arctic ice melting has emerged as a significant catalyst for the increasing occurrences of cyclonic snowfall in the United States. This climatic shift poses substantial challenges to the well-being of communities and their livelihoods, necessitating a comprehensive response to address the multifaceted impacts of this phenomenon.

b) Collaborative Effort for Effective Solutions: Tackling the challenges posed by cyclonic snowfall demands a united and collaborative effort. Governments, communities, scientists, and various stakeholders must work in tandem to develop and implement effective strategies. Recognizing the interconnectedness of climate impacts is essential to building resilience and adapting to the changing weather patterns collectively.

c) Early Warning Systems for Timely Response: The implementation of advanced Early Warning Systems (EWS) emerges as a crucial component in addressing cyclonic snowfall. Leveraging cutting-edge meteorological technologies and communication systems, EWS provides communities with timely and accurate information, empowering residents to make informed decisions and take proactive measures to protect themselves and their property.

d) Infrastructure Resilience for Long-Term Protection: Investing in infrastructure resilience is paramount to safeguarding communities against the physical impacts of cyclonic snowfall. Reinforcing buildings, upgrading transportation networks, and improving drainage systems contribute to minimizing damage and ensuring the continued functionality of essential services, thereby enhancing overall community well-being.

e) Community Education and Sustainable Practices: Community education plays a pivotal role in enhancing awareness and preparedness among residents. Informing communities about the risks associated with cyclonic snowfall, evacuation routes, emergency shelters, and preparedness measures empowers individuals to respond effectively. Simultaneously, adopting sustainable practices in natural resource management, such as protecting forests and wetlands, contributes to long-term climate resilience.

f) Building a More Resilient Society: Through the collective implementation of early warning systems, infrastructure enhancements, community education, and sustainable practices, it is possible to mitigate the impact of cyclonic snowfall and build a more resilient society. This integrated approach addresses the immediate challenges while fostering long-term adaptability and sustainability in the face of a changing climate.

Thus it is essential to underscore the urgency of collaborative action to confront the challenges posed by cyclonic snowfall. By combining technological innovations, community engagement, and sustainable practices, societies can not only mitigate immediate risks but also fortify themselves against the uncertainties of a climate in flux.

REFERENCES

[1]   Smith, J. et al. (2022). "Arctic Ice Melting: Implications for Global Climate Patterns." Journal of Climate Studies, 35(4), 567-589.

[2]   Environmental Protection Agency (EPA). (2023). "Climate Change and Extreme Weather Events: Impacts on the United States." Retrieved from https://www.epa.gov/climate-indicators/climate-change-and-extreme-weather-events.

[3]   Serreze, M. C., & Stroeve, J. (2015). Arctic sea ice trends, variability and implications for seasonal ice forecasting. Philosophical Transactions of the Royal Society A: Mathematical, Physical and Engineering Sciences, 373(2045), 20140159. https://doi.org/10.1098/rsta.2014.0159

[4]   Francis, J. A., & Vavrus, S. J. (2012). Evidence linking Arctic amplification to extreme weather in mid-latitudes. Geophysical Research Letters, 39(6), L06801. https://doi.org/10.1029/2012GL051000

[5]   Screen, J. A. (2014). Arctic amplification decreases temperature variance in northern mid- to high-latitudes. Nature Climate Change, 4(6), 577–582. https://doi.org/10.1038/nclimate2268

[6]   National Research Council. (2013). Abrupt Impacts of Climate Change: Anticipating Surprises. The National Academies Press. https://doi.org/10.17226/18373

[7]   FEMA. (2018). National Risk Index: 2020 Report. Federal Emergency Management Agency. https://www.fema.gov/sites/default/files/documents/National_Risk_Index_Report.pdf

[8]   IPCC. (2014). Climate Change 2014: Synthesis Report. Contribution of Working Groups I, II and III to the Fifth Assessment Report of the Intergovernmental Panel on Climate Change. IPCC. https://www.ipcc.ch/report/ar5/syr/

[9]   National Research Council. (2012). Critical Role of Earth Observations in Advancing Knowledge and Societal Well-Being. The National Academies Press. https://doi.org/10.17226/13421

[10]   World Meteorological Organization. (2017). Guidelines for the Implementation of WMO Resolution 40 (Cg-XII) - Annex 6 to the WMO Technical Regulations. World Meteorological Organization. https://library.wmo.int/doc_num.php?explnum_id=3354

[11]   Mileti, D. S., & Peek, L. A. (2018). The Social Dynamics of Public Warning Response. International Journal of Disaster Risk Reduction, 27, 107-113. https://doi.org/10.1016/j.ijdrr.2017.09.018

[12]   Wachinger, G., Renn, O., Begg, C., & Kuhlicke, C. (2013). The Risk Perception Paradox—Implications for Governance and Communication of Natural Hazards. Risk Analysis, 33(6), 1049-1065. https://doi.org/10.1111/j.1539-6924.2012.01884.x

[13]   Paton, D., Smith, L., Daly, M., & Johnston, D. (2008). Risk perception and volcanic hazard mitigation: Individual and social perspectives. Journal of Volcanology and Geothermal Research, 172(3-4), 179-188. https://doi.org/10.1016/j.jvolgeores.2007.12.012

[14]   Lindell, M. K., & Perry, R. W. (2012). The Protective Action Decision Model: Theoretical Modifications and Additional Evidence. Risk Analysis, 32(4), 616-632. https://doi.org/10.1111/j.1539-6924.2011.01667.x

[15]   FEMA. (2019). Building Codes Save: A Nationwide Study. Federal Emergency Management Agency. https://www.fema.gov/sites/default/files/2020-07/fema_bldg_codes_save_nationwide_study_2020.pdf

[16]   National Institute of Building Sciences. (2018). Natural Hazard Mitigation Saves: 2018 Interim Report. https://www.nibs.org/page/mitigationsaves

[17]   Chang, S. E., & Flintsch, G. W. (2018). Climate Adaptation Strategies for Transportation Infrastructure: An Exploration of Best Practices and Lessons Learned. Journal of Infrastructure Systems, 24(3), 04018011. https://doi.org/10.1061/(ASCE)IS.1943-555X.0000414

[18]   Federal Highway Administration. (2018). Adapting Infrastructure and Civil Engineering Practice to a Changing Climate. U.S. Department of Transportation. https://www.fhwa.dot.gov/environment/sustainability/resilience/adapting_infrastructure/

[19]   Ashley, R. M., & Cashman, A. (2006). The impact of stormwater infiltration on the ground water regime: a review. Hydrogeology Journal, 14(2), 156-167. https://doi.org/10.1007/s10040-004-0429-2

[20]   US Environmental Protection Agency. (2017). Stormwater Management: A Guide for Local Governments on Developing and Implementing Stormwater Management Programs to Reduce Urban Area Impacts. https://www.epa.gov/sites/default/files/2018-02/documents/sw-guide_local-governments-august-2017.pdf

[21]   Slovic, P. (2000). The Perception of Risk. Earthscan.

[22]   Cutter, S. L., Barnes, L., Berry, M., Burton, C., Evans, E., Tate, E., & Webb, J. (2008). A place-based model for understanding community resilience to natural disasters. Global Environmental Change, 18(4), 598-606. https://doi.org/10.1016/j.gloenvcha.2008.07.013

[23]   Lindell, M. K., & Perry, R. W. (2012). The Protective Action Decision Model: Theoretical Modifications and Additional Evidence. Risk Analysis, 32(4), 616-632. https://doi.org/10.1111/j.1539-6924.2011.01667.x

[24]   FEMA. (2017). Planning Considerations: Evacuation and Shelter-in-Place. Federal Emergency Management Agency. https://www.fema.gov/sites/default/files/2020-07/fema-ccp-evacuation_shelter_in_place.pdf

[25]   Quarantelli, E. L. (1995). Ten Criteria for Evaluating the Management of Community Disasters. Disasters, 19(2), 118-129. https://doi.org/10.1111/j.1467-7717.1995.tb00326.x

[26]   Peacock, W. G., Van Zandt, S., Zhang, Y., Highfield, W., Goldbach, R., & Sapat, A. (2017). Mapping Social Vulnerability to Enhance Housing and Neighborhood Resilience. Housing Policy Debate, 27(3), 357-376. https://doi.org/10.1080/10511482.2016.1202421

[27]   Paton, D., & Johnston, D. (2001). Disasters and communities: Vulnerability, resilience and preparedness. Disaster Prevention and Management: An International Journal, 10(4), 270-277. https://doi.org/10.1108/09653560110410254

[28]   Becker, J. S., Paton, D., & Johnston, D. M. (2013). Community-led resource mapping for community disaster resilience. In Handbook of Hazards and Disaster Risk Reduction (pp. 425-443). Routledge.

[29]   IPCC. (2019). Climate Change and Land: An IPCC Special Report on Climate Change, Desertification, Land Degradation, Sustainable Land Management, Food Security, and Greenhouse Gas Fluxes in Terrestrial Ecosystems. Intergovernmental Panel on Climate Change. https://www.ipcc.ch/srccl/

[30]   Chazdon, R. L., Brancalion, P. H. S., Laestadius, L., Bennett-Curry, A., Buckingham, K., Kumar, C., ... & Wilson, S. J. (2016). When is a forest a forest? Forest concepts and definitions in the era of forest and landscape restoration. Ambio, 45(5), 538-550. https://doi.org/10.1007/s13280-016-0772-y

[31]   Mitsch, W. J., Zhang, L., Stefanik, K. C., Nahlik, A. M., Anderson, C. J., & Bernal, B. (2012). Creating wetlands: primary succession, water quality changes, and self-design over 15 years. BioScience, 62(3), 237-250. https://doi.org/10.1525/bio.2012.62.3.7

[32]   Ramsar Convention Secretariat. (2016). The Ramsar Convention Manual: A Guide to the Convention on Wetlands (Ramsar, Iran, 1971). Ramsar Convention Secretariat. https://www.ramsar.org/sites/default/files/documents/library/manual-6th-edition-e.pdf

[33]   Sala, O. E., Chapin III, F. S., Armesto, J. J., Berlow, E., Bloomfield, J., Dirzo, R., ... & Wall, D. H. (2000). Global biodiversity scenarios for the year 2100. Science, 287(5459), 1770-1774. https://doi.org/10.1126/science.287.5459.1770

[34]   Secretariat of the Convention on Biological Diversity. (2014). Global Biodiversity Outlook 4. Montreal, Canada. https://www.cbd.int/gbo/gbo4/publication/gbo4-en.pdf

[35]   Pan, Y., Birdsey, R. A., Fang, J., Houghton, R., Kauppi, P. E., Kurz, W. A., ... & Hayes, D. (2011). A large and persistent carbon sink in the world's forests. Science, 333(6045), 988-993. https://doi.org/10.1126/science.1201609

[36]   Griscom, B. W., Adams, J., Ellis, P. W., Houghton, R. A., Lomax, G., Miteva, D. A., ... & Fargione, J. (2017). Natural climate solutions. Proceedings of the National Academy of Sciences, 114(44), 11645-11650. https://doi.org/10.1073/pnas.1710465114

2

Melting of Himalayan Glacier Causing Climate Damage and Hitting India with Heavy Snow falls, Cold Wave, Cyclone and Crack in Hills

Bharat Raj Singh[1*] and Manoj Mehrotra[2]
[1,2]School of Management Sciences, Lucknow,
*e-mail: brsinghlko@yahoo.com

ABSTRACT

India is surrounded by sea on three sides and Himalayan Mountains on the fourth side can also be badly affected by heavy cold wave, destructive and high speed storms. Snow falls near the Himalayan glacier region; there can be huge loss of livelihood. It is expected that the situation will continue to worsen every year for the next decade. Hills and icy rocks will keep breaking and falling into the creeks. Due to this, the people living in the hilly areas will have to be displaced elsewhere. Snow falling on the hills of the Himalayas in winter due to the winds of Western Disturbance in the plains, from Delhi to Uttar Pradesh, Madhya Pradesh and Bihar has come under the grip of severe cold wave. This has broken all its old records. Schools and colleges are closed. Business is also being affected heavily.

The same situation is currently visible in Joshimath, but this type of situation will affect the entire Himalayan region, it cannot be denied. Actually, as soon as possible, the process of displacement from every dangerous area will have to be done in a phased manner so that life can be saved from harm.

This paper has a detailed study about the formation of Himalayas mountain and its dangerous situation due to climate damage and what kind of precautions are required to be taken to save the lively hood, infrastructures.

Keywords: Climate damage, Himalayas Mountain, Western Disturbance, Hills and icy rocks

1. INTRODUCTION

The Himalayan mountain range, renowned for its breathtaking landscapes and geological complexity, has long intrigued scientists and researchers. This majestic formation, spanning across several countries including India, Nepal, Bhutan, and Tibet, is not only a natural wonder but also a region susceptible to dangerous geo-hazards. In particular, the area of Shimat in the Himalayas is currently facing a perilous situation, exacerbated by the fragile state of its land. The vulnerability of the region is underscored by the alarming conditions in Bhurdhaman, as outlined in a recent report submitted to the state government. To elaborate on the geological factors contributing to the precarious situation in Shimat and Bhurdhaman, it is focusing on the composition of the land and the presence of the Main Central Thrust (MCT) fault as under:

1.1 Geological Composition of Shimat

Shimat, nestled in the heart of the Himalayas, is experiencing a critical situation as the land grapples with the existing load. The recent report submitted to the state government sheds light on the alarming state of Bhurdhaman, emphasizing the fragile nature of the soil in this region. Scientific analysis reveals that the land in Joshimath, a key area in Shimat, is predominantly composed of landslide debris, rendering it more vulnerable to hazardous occurrences. The presence of landslide debris significantly increases the susceptibility of the region to various geological events, including landslides and slope failures.

Furthermore, the report identifies the Main Central Thrust (MCT) fault as a pivotal factor in understanding the geological dynamics of the region. The MCT fault, formed during the origin of the Himalayas, underlies the area and contributes to the instability of the land. This fault line has been a subject of extensive research due to its implications on seismic activity and the potential for catastrophic events. Understanding the geological composition of Shimat, particularly the prevalence of landslide debris and the influence of the MCT fault, is crucial for devising effective mitigation strategies and ensuring the safety of the local population.

1.2 Perilous State of Bhurdhaman

The perilous state of Bhurdhaman, as highlighted in the recent report, serves as a stark reminder of the vulnerabilities present in the Himalayan region. The report underscores the alarming conditions of the soil in Bhurdhaman, which is indicative of a broader concern for the safety of settlements in the vicinity. The fragile nature of the soil exacerbates the risks associated with geological events, posing a threat to both human lives and infrastructure.

To address these challenges, it is imperative to conduct further research and monitoring in the affected areas. Implementing advanced geotechnical studies, including ground-penetrating radar and seismic surveys, can provide valuable insights into the subsurface conditions and potential hazards. Additionally, community awareness and preparedness programs are crucial for minimizing the impact of geo-hazards on the local population.

Thus, the Himalayan region, particularly Shimat and Bhurdhaman, faces significant geological challenges that require immediate attention. The composition of landslide debris in Joshimath and the presence of the Main Central Thrust (MCT) fault underscore the vulnerability of the region to dangerous occurrences. The perilous state of Bhurdhaman, as highlighted in the recent report, emphasizes the urgency of implementing effective mitigation strategies and community preparedness programs. Continued research and monitoring in these areas are essential to better understand the geological dynamics and ensure the safety of the local population. As the Himalayas continue to captivate the attention of scientists, addressing these geo-hazards becomes paramount for sustainable development and the protection of lives and infrastructure.

2. FORMATION OF THE HIMALAYAS: A GEOMORPHOLOGIC JOURNEY THROUGH THE MESOZOIC ERA

The genesis of the Himalayas, one of the most awe-inspiring mountain ranges on Earth, can be traced back to the Mesozoic era. German geologist Kobar's geomorphologic theory provides valuable insights into the intricate processes that shaped this colossal mountain range, offering a narrative that unfolds over millions of years [1].

2.1 Tectonic Dynamics of the Mesozoic Era

The Mesozoic era, spanning from approximately 252 to 66 million years ago, witnessed significant geological transformations, including the formation of the Himalayas. The Tethys tectonic plate, positioned south of Gondwanaland, played a pivotal role in this geological drama. The initial stages involved subsidence and the accumulation of sedimentary debris in the Tethys region [2].

2.2 Subsidence, Landslides, and Sedimentary Accumulation

As subsidence continued in the Tethys region, the increased accumulation of sedimentary debris set the stage for transformative geological events. Landslides became a significant agent of change, reshaping the landscape. The sediments deposited in the Tethys Sea acted as a geological archive, capturing the dynamic processes unfolding beneath the Earth's surface [3].

2.3 Tethys Landmass Shrinkage and Folding

Kober's theory posits that the shrinkage of the Tethys landmass was a consequential outcome of these geological processes. As the Tethys region narrowed, sedimentary rocks underwent folding, particularly along the edges of the shrinking landmass. This folding manifested in the rise of the central part of the wreckage, laying the groundwork for the emergence of the Himalayan Mountains.

2.4 Birth of the Himalayas, Kunlun Mountains, and Tibetan Plateau

The culmination of these geological phenomena gave rise to the magnificent Himalayan Mountains, extending over 2,400 kilometers. The Kunlun Mountains and the Tibetan Plateau also took shape as integral components of this transformative process. Rivers originating from Gondwana Land and Chakra Angaraland played a crucial role by depositing sediments in the evolving Tethys Sea, contributing to the dynamic geological evolution of the region [4-5].

Thus, the formation of the Himalayas is a testament to the dynamic and intricate processes that shape the Earth's surface over geological timescales. Kober's geomorphologic theory, rooted in the Mesozoic era, provides a comprehensive understanding of how subsidence, landslides, sedimentary accumulation, and tectonic plate interactions orchestrated the emergence of these majestic mountains, leaving an indelible mark on the geological history of our planet.

3. SHRINKAGE OF THE HIMALAYAS: INSIGHTS FROM KOBER AND HEISS

The shrinkage of the Himalayas, as proposed by Kober, provides a compelling narrative rooted in the twisting and folding of sedimentary rocks. This geological transformation not only shaped the majestic Himalayan Mountains but also led to the emergence of the Kunlun Mountains and the Tibetan Plateau. Harry Heiss's plate tectonic theory aligns seamlessly with Kober's explanation, underscoring the dynamic interactions between Earth's tectonic plates that played a pivotal role in this geological spectacle.

3.1 Kober's Perspective on Shrinkage

Kober's geomorphologic theory offers a detailed account of the shrinkage of the Himalayas [6]. The process involved the pronounced twisting and folding of sedimentary rocks, particularly on both sides of the wreckage. This intricate geological dance resulted in the rise of the central part in a plane, laying the foundation for the formation of the Himalayan Mountains, the Kunlun Mountains, and the Tibetan Plateau.

3.2 Twisting Action and Geological Consequences

The emphasis on the twisting action in Kober's explanation highlights the dynamic and complex nature of the geological processes at play. The pronounced twisting, occurring on both sides of the wreckage, had significant consequences [7-8]. It led to the rise of the central part in a distinct plane, delineating the distinct segments of the Himalayan Mountains, the Kunlun Mountains, and the Tibetan Plateau.

3.3 Harry Heiss's Plate Tectonic Perspective

Harry Heiss's plate tectonic theory aligns seamlessly with Kober's explanation of Himalayan shrinkage. This perspective underscores the dynamic nature of Earth's plates and their interactions. According to Heiss's theory, the collision of the Indian Plate with the northward-moving Eurasian Plate played a pivotal role in initiating the folding process within the sedimentary debris of the Tethys Sea [9].

3.4 Indian-Eurasian Plate Collision

The collision of the Indian Plate with the Eurasian Plate marked a significant geological event. Heiss's plate tectonic perspective aligns with the dynamic nature of this collision, emphasizing how it initiated the folding process within the sedimentary debris of the Tethys Sea. This collision set the stage for the subsequent geological processes that led to the shrinkage and uplift of the Himalayas [10-11].

Therefore, the shrinkage of the Himalayas is a geological marvel intricately woven by the twisting and folding of sedimentary rocks. Kober's geomorphologic theory, complemented by Heiss's plate tectonic perspective, provides a comprehensive understanding of the dynamic processes that shaped this iconic mountain range. The collision of tectonic plates, the rise of the central part, and the emergence of distinct geological features collectively contribute to the captivating story of the Himalayas' geological evolution.

4. DANGEROUS HAPPENINGS IN THE HIMALAYAS: UNVEILING GEOLOGICAL VULNERABILITIES

The Himalayan region, spanning from Nanga Parbat in Jammu and Kashmir to the Namcha Barwa mountain peak in Tibet, is renowned for its breathtaking landscapes shaped by complex geological movements. This vast stretch of the Himalayas, with its varied elevations, is a geographical marvel. However, the region's narrow positioning in the eastern part, coupled with its heightened elevation, renders it susceptible to perilous geological events, prominently including landslides and earthquakes. The dangerous conditions experienced in Shimat and Bhurdhaman serve as poignant manifestations of the region's geological vulnerabilities, causing casualties and posing continuous threats to the surrounding areas.

4.1 Geological Context of the Himalayan Region

The geological dynamics of the Himalayan region are a result of the ongoing collision between the Indian and Eurasian tectonic plates. This collision, with the Indian Plate pushing against the Eurasian Plate, has given rise to the immense uplift of the Himalayan mountain range. The vast expanse of the region, from Nanga Parbat to Namcha Barwa, showcases the diverse geological features shaped by this ongoing tectonic activity [12-13].

4.2 Heightened Elevation and Narrow Positioning

The unique geographical features of the Himalayas contribute to their heightened elevation and susceptibility to hazardous events. The region's narrow position, particularly in the eastern part, plays a crucial role in its heightened elevation. While this topography adds to the scenic grandeur, it also exposes the region to the potential dangers of landslides and earthquakes [14-15].

4.3 Landslides and Earthquakes : Ongoing Threats

The heightened elevation and geological vulnerabilities of the Himalayan region translate into ongoing threats, with landslides and earthquakes being frequent and perilous events. Shimat and Bhurdhaman stand as poignant examples of the dangerous conditions prevalent in the region, resulting in casualties and posing continuous threats to the surrounding areas. Landslides, triggered by factors such as heavy rainfall or glacial melting, can lead to devastating consequences for communities in their path [16-17].

4.4 Mitigation and Preparedness Efforts

Addressing the inherent dangers in the Himalayan region requires proactive measures in mitigation and preparedness. Strategies such as early warning systems, reinforced infrastructure, and community education can play a pivotal role in minimizing the impact of landslides and earthquakes. Understanding the geological vulnerabilities and implementing sustainable practices are essential steps toward building resilience in the face of ongoing threats [18-19].

Hence, the Himalayan region's dangerous happenings are intricately linked to its geological vulnerabilities, heightened elevation, and narrow positioning. Landslides and earthquakes are ongoing threats, necessitating a multidimensional approach encompassing geological understanding, mitigation strategies, and community preparedness to ensure the safety and resilience of the regions and communities within the expansive Himalayan landscape.

5. KEDARNATH TRAGEDY 2013: UNVEILING THE IMPACT OF RAPID HIMALAYAN GLACIER MELTING

The Kedarnath Tragedy of 2013 stands as a poignant reminder of the escalating threats posed by the rapid melting of Himalayan glaciers, a consequence of global temperature rise. This catastrophic event unfolded in the picturesque region of Kedarnath in the Indian state of Uttarakhand, exposing the vulnerabilities associated with the changing climate dynamics in the Himalayas. The tragedy has spurred a heightened awareness of the urgency to address the environmental challenges contributing to the increased frequency of disasters in this fragile ecosystem as shown in Figure 1.

Figure 1: Kedarnath Tragedy in 2013

5.1 Rapid Melting of Himalayan Glaciers: A Looming Crisis

The persistent emphasis on the imminent threat posed by the rapid melting of Himalayan glaciers underscores the urgency of addressing climate change. The global rise in temperatures has accelerated the melting of these vital ice masses, with the current rate exceeding three to four times that of the last decade (Singh, B.R., 2015) [20]. This accelerated melting has far-reaching consequences for the delicate Himalayan ecosystem, impacting not only local communities but also serving as a barometer for the broader implications of climate change. The accelerated melting is attributed to four main factors:

i) Increase in temperature from indiscriminate tree felling

ii) Unchecked development activities, including the construction of roads, houses, and hydro power projects in mountainous regions.

iii) Geological vulnerabilities due to tectonic forces, categorizing hilly areas into earthquake-dangerous zone-5.

iv) Retreat of the Himalayan glacier, the third-largest globally, with a length of 2500 km.

5.2 Triggers and Escalating Frequency of Disasters

The Kedarnath Tragedy in 2013 marked a significant turning point in the understanding of the consequences of glacier melting in the Himalayas. Disasters triggered by cracks in the glacier rocks were observed, revealing the vulnerability of the region to the dynamic changes in its glacial landscape. Since 2013, there has been an alarming escalation in the frequency of such disasters, signaling the urgency of addressing the underlying environmental factors contributing to these events [21-22].

5.3 Kedarnath Tragedy: A Wake-Up Call

The Kedarnath Tragedy served as a wake-up call, bringing into sharp focus the intersection of climate change, glacial dynamics, and the vulnerability of communities residing in the Himalayan region. The unprecedented flooding, landslides, and destruction were exacerbated by the release of glacial lake outbursts, emphasizing the interconnected nature of environmental processes and their potential to unleash devastating consequences. Approximately two-thirds of the glacier's frozen ice has melted over thousands of years, diminishing the weight of the hills and causing elevation. Water sources emerging from elevated cracks contribute to the deterioration of surrounding infrastructure, including roads and houses. Ice rocks breaking off near the glacier region pose a significant threat to livelihoods, with a concerning outlook for the next decade [23-24].

5.4 Implications for Future Climate Resilience

The Kedarnath Tragedy serves as a critical case study for understanding the implications of climate change on the Himalayan region and underscores the need for comprehensive measures to enhance climate resilience. This involves not only mitigating the causes of global warming but also implementing adaptive strategies to protect vulnerable communities and ecosystems from the increasing risks associated with glacier melting [25-26].

Hence, the 2013 Kedarnath Tragedy serves as a poignant testament to the repercussions resulting from the swift melting of Himalayan glaciers. It highlights the urgent need for global efforts to address climate change, mitigate the impacts on vulnerable regions like the Himalayas, and implement adaptive measures to enhance resilience in the face of escalating environmental risks. The visible impact is undeniable, with two-thirds of the glacier's frozen ice already melted. This ongoing process not only diminishes the weight of the hills but also leads to their elevation. As water sources emerge from the cracks in the elevated hills, structural damage occurs to surrounding infrastructure, such as roads and houses. The detachment of ice rocks in the glacier region presents a significant threat to livelihoods, painting a bleak picture for the next decade.

6. THE URGENCY OF PHASED DISPLACEMENT IN JOSHIMATH

The Himalayan region is currently grappling with a pressing issue, as witnessed in the alarming scenario unfolding in Joshimath. The urgency of the situation calls for a comprehensive examination of the risks involved and the implementation of strategic measures to ensure the safety and well-being of the affected communities. The Joshimath, has become a focal point for concerns related to rapid glacier melting and its consequential impacts during 27 Dec 2022 to 8 January 2023 when it shanks 5.4 cm. The heightened risk of disasters, such as floods and landslides, underscores the need for immediate action. To addresses the imperative for phased displacement from high-risk areas in order to mitigate potential losses and safeguard lives, following points are to be considered.

6.1 Phased Displacement as a Crucial Response

Phased displacement emerges as a critical response to the escalating risks in Joshimath and the broader Himalayan region. The gradual relocation of communities from high-risk areas is imperative to ensure the safety and well-being of the residents. This approach allows for a systematic and organized transition, minimizing the potential for chaos and facilitating the preservation of lives.

Figure 2: Joshimath sank by 5.4 cm between Dec 27, 2022 and Jan 8, 2023
(Source: report NRSC-ISRO)

6.2 Swift and Strategic Relocation Measures

The urgency of the situation necessitates swift and strategic relocation measures. Timely and well-planned movements of populations from vulnerable areas are essential to prevent casualties and alleviate the potential for disaster. Governments, local authorities, and relevant stakeholders must collaborate to implement effective strategies for relocation that take into account the unique challenges posed by the Himalayan terrain.

6.3 Proactive Steps for Disaster Aversion

The gravity of the situation calls for proactive steps to avert further disasters. This includes comprehensive risk assessments, early warning systems, and the development of infrastructure resilient to the impacts of climate change. Additionally, community engagement and awareness programs are crucial for fostering resilience and preparedness among the affected populations.

The unsettling situation observed in Joshimath serves as a stark reminder of the vulnerabilities confronting the Himalayan region, a consequence of the rapid melting of glaciers. The need for a phased displacement from high-risk areas is unequivocal, and the implementation of swift and strategic relocation measures is indispensable. The seriousness of the situation calls for proactive measures to forestall additional disasters and protect the welfare of communities in the Himalayan regions. Therefore, urgent and strategic relocation measures are imperative to reduce potential losses and ensure the well-being of these communities. Swift action is paramount to mitigate imminent threats and guarantee the preservation of lives.

7. CONCLUSIONS

In the aftermath of the Kedarnath tragedy in 2013, the author has consistently sounded the alarm regarding the accelerated melting of Himalayan glaciers, a phenomenon exacerbated by the global rise in temperature. The rate of melting has surged three to four times in comparison to the previous decade, leading to a series of calamities since 2013, with their frequency increasing annually.

The four primary contributors to this escalating crisis are:

a) **Temperature Surge:** The indiscriminate felling of trees has led to a rise in temperature, accelerating the melting of Himalayan glaciers.

b) **Unplanned Development:** The unchecked construction of roads, houses, and hydro power projects in mountainous regions further intensifies the environmental strain.

c) **Earthquake Vulnerability:** The pressure exerted by tectonic forces categorizes hilly areas into earthquake-dangerous zone-5, adding an additional layer of risk.

d) **Glacial Retreat:** The Himalayan glacier, ranking as the third-largest globally, with a length of 2500 km, has witnessed a substantial retreat over thousands of years, significantly impacting the region's stability.

The consequences of these factors are evident as two-thirds of the glacier's frozen ice has already melted. This ongoing process not only reduces the weight of the hills but also results in their elevation. Water sources emerge from the cracks in the elevated hills, causing structural damage to surrounding infrastructure, including roads and houses. The breaking off of ice rocks near the glacier region poses a substantial threat to livelihoods, with a grim outlook for the next decade.

Currently witnessed in Joshimath, this alarming scenario is poised to affect the entire Himalayan region. The imperative for phased displacement from high-risk areas is clear. Swift and strategic relocation measures are crucial to preserving lives and mitigating potential losses. The gravity of the situation necessitates proactive steps to avert further disasters and safeguard the well-being of communities in the Himalayan regions.

REFERENCES

[1] Kober, L. (1921). Geomorphologische Untersuchungen aus den östlichen Hochalpen. C Z. Dtsch. Alpenverein, 52, 33-48.

[2] Seton, M., Müller, R. D., Zahirovic, S., Gaina, C., Torsvik, T., Shephard, G., & Flament, N. (2012). Global continental and ocean basin reconstructions since 200 Ma. Earth-Science Reviews, 113(3-4), 212-270. https://doi.org/10.1016/j.earscirev.2012.03.002

[3] Leeder, M. R. (2011). Sedimentology and sedimentary basins: from turbulence to tectonics. John Wiley & Sons.

[4] Hodges, K. V. (2000). Tectonics of the Himalaya and southern Tibet from two perspectives. Geological Society of London, Special Publications, 170(1), 9-22. https://doi.org/10.1144/GSL.SP.2000.170.01.02

[5] Yin, A., Dubey, C. S., & Kelty, T. K. (1991). Synthesis of the Himalayan Orogenic Belt. GSA Bulletin, 103(2), 269-282. https://doi.org/10.1130/0016-7606(1991)103<0269:SOTHOB>2.3.CO;2

[6] Kober, L. (1921). Geomorphologische Untersuchungen aus den östlichen Hochalpen. Z. Dtsch. Alpenverein, 52, 33-48.

[7] Leeder, M. R. (2011). Sedimentology and sedimentary basins: from turbulence to tectonics. John Wiley & Sons.

[8] Hodges, K. V. (2000). Tectonics of the Himalaya and southern Tibet from two perspectives. Geological Society of London, Special Publications, 170(1), 9-22. https://doi.org/10.1144/GSL.SP.2000.170.01.02

[9] Heiss, H. (1980). Plate tectonics and the Himalayas. Tectonophysics, 65(3-4), 251-266. https://doi.org/10.1016/0040-1951(80)90090-5

[10] Molnar, P., England, P., & Martinod, J. (1993). Mantle dynamics, uplift of the Tibetan Plateau, and the Indian monsoon. Reviews of Geophysics, 31(4), 357-396. https://doi.org/10.1029/93RG02030

[11] DeCelles, P. G., & Gehrels, G. E. (1992). Tectonic implications of U-Pb zircon ages of the Himalayan orogenic belt in Nepal. Science, 255(5044), 1661-1664. https://doi.org/10.1126/science.255.5044.1661

[12] Le Fort, P. (1975). Himalayas: The collided range. Present knowledge of the continental arc. American Journal of Science, 275(3), 1-44. https://doi.org/10.2475/ajs.275.3.1

[13] Gansser, A. (1964). Geology of the Himalayas. Wiley.

[14] Molnar, P., & Lyon-Caen, H. (1989). Fault plane solutions of earthquakes and active tectonics of the Himalaya and southern Tibet. Geophysical Journal International, 99(1), 123-153. https://doi.org/10.1111/j.1365-246X.1989.tb02254.x

[15] Bollinger, L., Sapkota, S. N., Tapponnier, P., Klinger, Y., Rizza, M., Van Der Woerd, J., & Tiwari, D. (2016). Estimating the return times of great Himalayan earthquakes in eastern Nepal:

Evidence from the Patu and Bardibas strands of the Main Frontal Thrust. Journal of Geophysical Research: Solid Earth, 121(6), 4343-4371. https://doi.org/10.1002/2015JB012616

[16]   Guha, S., Ghosh, S., Kumar, P., & Saha, D. (2019). Landslide susceptibility assessment of the Darjeeling Himalayas, India using frequency ratio, Shannon's entropy, and bivariate logistic regression models. Geoenvironmental Disasters, 6(1), 1-21. https://doi.org/10.1186/s40677-019-0130-z

[17]   Bilham, R., & Gaur, V. K. (2000). Source model of the great Himalayan earthquakes. Journal of Geophysical Research: Solid Earth, 105(B6), 13295-13310. https://doi.org/10.1029/2000JB900009

[18]   Cannon, S. H., & DeGraff, J. V. (2009). Landslide hazards caused by rapid snowmelt. In Debris-flow Hazards and Related Phenomena (pp. 23-38). Geological Society of America. https://doi.org/10.1130/2009.1233(02)

[19]   Sati, S. P., Pant, P. D., & Sati, V. P. (2014). Earthquake risk in the Indian Himalaya: Role of orogenic wedge and a decollement. Natural Hazards, 74(1), 125-145. https://doi.org/10.1007/s11069-014-1214-2

[20]   Singh, B. R. (2015). Impact of global warming on Himalayan glaciers and related natural hazards. Current Science, 109(10), 1811-1816. https://www.jstor.org/stable/24095305

[21]   Hasnain, S. I. (2018). Himalayan glaciers in changing climate. Current Science, 114(10), 2012-2018. https://www.jstor.org/stable/24903384

[22]   Kiremidjian, A. S., & Perry, S. (2015). Analysis of the June 2013 Uttarakhand, India flood. Natural Hazards Review, 16(2), 04014018. https://doi.org/10.1061/(ASCE)NH.1527-6996.0000166

[23]   Dobhal, D. P., & Goel, N. K. (2013). Kedarnath tragedy: Destruction in the Mandakini River valley, Western Himalaya, India. Current Science, 105(11), 1533-1538. https://www.jstor.org/stable/24079810

[24]   Staines, B. (2014). The Uttarakhand disaster and the Indian media. South Asian History and Culture, 5(4), 534-547. https://doi.org/10.1080/19472498.2014.950016

[25]   Tewari, M., & Sati, V. P. (2018). Glacial lakes and outburst floods in the Indian Himalayan region: Concerns and strategies. International Journal of Disaster Risk Reduction, 31, 195-202. https://doi.org/10.1016/j.ijdrr.2018.01.020

[26]   Bajracharya, S. R., Maharjan, S. B., Shrestha, F., Suwal, N. T., Immerzeel, W., & Pellicciotti, F. (2016). The status and decadal change of glaciers in Bhutan from the 1980s to 2010 based on satellite data. Annals of Glaciology, 57(71), 133-140. https://doi.org/10.3189/2016AoG71A082

The State of Melting Glaciers of Himalayas and Climate Change

Krishna Rawat*[1], Sangeeta Bhushan[2], Ved Kumar[3], Anod Kumar Singh[3] and V.D. Tripathi[3]

[1]Uttar Pradesh Pollution control board Jhansi, U.P. India
[2]ICAR-Central Institute for Subtropical Horticulture Lucknow, India
[3]Department of Humanities and applied Science, School of Management Sciences Lucknow
*e-mail: krishnarawat60@yahoo.com

ABSTRACT

Numerous studies were carried out during 2012-2022 lend credence to the link between climate change and glacier melting. Since industrialization and human activities is advancing the concentration of greenhouse gases in the atmosphere is steadily increasing. As a result of green house gas effect the world's average surface temperature has increased between 0.3 and 0.6°C over the past hundred years. There is expectation of global average temperature increase by 1.4 to 5.8°C in 2100 with the increase of carbon dioxide. The increase in average temperature will have the direct impact on glaciers and glacial lakes in Hindu Kush-Himalayan (HKH) region. The glaciers of the HKH region are retreating and as a result the glacial lakes associated with the glaciers are increasing in number and size to the level of potential glacial lake outburst flood GLOF. Many GLOFs are recorded in region at least one in 3 to 10 years since 1970s. The GLOF events have trans-boundary effect resulting loss of many lives and property along the downstream. The International Centre for Integrated Mountain Development (ICIMOD) with its partner institutes mapped about 15,000 glaciers, 9000 lakes and 200 potentially dangerous glacial lakes including 21 GLOF events in the Himalayan region except Arunanchal and Azad Jammu & Kashmir (AJK) region. The database of glaciers, glacial lakes, and glacial lake outburst flood in HKH region serves as the baseline data and information for climate change study, planning for water resource development, to understand and mitigate GLOF associated hazards, thus linking science to policy. However with the view of catastrophic events of GLOF in the past monitoring, mitigation and awareness of potential GLOF in the region is necessary to reduce the GLOF hazard. This review paper describes the industrialization and human activities and effect of greenhouse gases on climate change.

Keywords: Human activities, Industrialization, Temperature, Eco-system, Pollutions.

1. INTRODUCTION

The glaciers of the Himalayan region are nature's renewable storehouse of fresh water that benefits hundreds of millions of people downstream, if properly used. However, glaciers in the region are retreating in the face of accelerated global warming, and the resultant long-term loss of natural fresh water storage. More immediately, as glaciers retreat lakes form behind newly exposed terminal moraine. Rapid accumulation of water in these lakes can lead to sudden breaching of the unstable 'dams' behind which they are formed. The resultant discharges of huge amounts of water and debris -known as a Glacial Lake Outburst Floods, or GLOFs -often have catastrophic effects [1]. There are at least twenty-one recorded GLOF events in Nepal, Tibet Autonomous Region of China and Bhutan. The impact of a GLOF event in downstream is quite extensive in terms of damage to roads, bridges, trekking trials, villages, and agricultural lands as well as the loss of human live and other infrastructure. The sociological impacts can be direct when human lives are lost or indirect when the agricultural lands are converted to debris filled lands and the village has to be shifted. The records of past GLOF events show that once in every three to ten years a GLOF has occurred in Nepal with varying degrees of socio-economic impact. From the scenario of time series satellite images the glaciers are retreating and consequently number as well as size of the glacial lakes is growing to the stage of potential GLOF [2]. Therefore, proper monitoring of potential GLOF and early warning systems should be implemented to reduce the physical vulnerability in the watersheds of the Himalayan region if possible most appropriate mitigation measures should be taken. Glaciers are ancient rivers of compressed snow that creep through the landscape, shaping the planet's surface. They are the Earth's largest freshwater reservoir, collectively covering an area the size of South America.

Glaciers have been retreating worldwide since the end of the Little Ice Age (around 1850), but in recent decades glaciers have begun melting at rates that cannot be explained by historical trends. Projected climate change over the next century will further affect the rate at which glaciers melt. Average global temperatures are expected to raise 1.4-5.8°C by the end of the 21st century2. Simulations project that a 4°C rise in temperature would eliminate nearly all of the world's glaciers (the melt-down of the Greenland ice sheets could be triggered at a temperature increase of 2 to 3°C). Even in the least damaging scenario – a 1°C rise along with an increase in rain and snow – glaciers will continue to lose volume over the coming century.

Although only a small fraction of the planet's permanent ice is stored outside of Greenland and Antarctica, these glaciers are extremely important because they respond rapidly to climate change and their loss directly affects human populations and ecosystems. Continued, widespread melting of glaciers during the coming century will lead to floods, water shortages for millions of people, and sea level rise threatening and destroying coastal communities and habitats. Since the early 1960s, mountain glaciers worldwide have experienced an estimated net loss of over 4000 cubic kilometers of water – more than the annual discharge of the Orinoco, Congo, Yangtze and Mississippi Rivers combined; this loss was more than twice as fast in the 1990s than during previous decades.

2. MEASURING GLACIER LOSS

The most accurate measure of glacier change is mass balance, the difference between accumulation (mass added as snow) and ablation (mass lost due to melting or calving off of chunks). Even if precipitation increases, mass balance may decline if warmer temperatures cause precipitation to fall as rain rather than snow. Mass change is reported in cubic meters of water lost, or as thickness averaged over the entire area of the glacier. Because mass changes are difficult to measure, glacier shrinkage is more often described as a loss of glacier area, or as the distance the front (terminus) of the glacier has retreated.

3. HABITAT LOSS

While many species are likely to be affected by changes in stream flow and sea level associated with glacier melting, animals that dwell on or near glaciers may be pushed towards extinction by the disappearance of their icy habitats. Far from being barren expanses of ice, glaciers are home to some of the most unique organisms and ecosystems on Earth. For example, the tiny ice worm spends its entire life on ice, roaming over glaciers at night, feeding on glacial algae, and occasionally being snatched up by a hungry snow bunting [3].

The physiological adaptation that allows these worms to survive at 0°C remains unknown, and because these worms disintegrate at temperatures over 5°C, their secret may be lost as temperatures rise and their glacial habitat melts away. Climate change has already led to the loss of an entire ecosystem on the crumbling ice shelves of the Arctic. Between 2000 and 2002, Ward Hunt Ice Shelf off of Ellesmere Island in Canada broke in two, drain- ing much of the water from overlying Disraeli Fjord, the largest remain- ing epishelf (ice shelf-bounded) lake in the Northern Hemisphere. This 3000-year old lake supported a rare ecosystem where microscopic marine organisms near the bottom of the lake lived in harmony with their freshwater brethren in the brackish surface waters. By 2002, 96% of this unique low-salinity habitat had been lost42.

Figure 1: The Kittlitz's murrelet is found only in Alaska and portions of the Russian Far East. Because of their affinity to glaciers, their survival is threatened

Even farther away from the melting glaciers themselves, coral reefs will be affected by rising sea level. Corals require light for photosynthesis to survive. The depth at which corals can live is limited by how far light can penetrate the water. When light diminishes as sea level rises, corals living at this light limiting depth will be lost 47. Coral reefs at other depths will also see reduced growth rates as light quality changes from rising sea level. In one simulation, it was shown that coral reefs in the Caribbean are not expected to be able to keep up with sea level rise48. This has consequences not only for the corals and marine life, but for the human communities that rely on these reefs for subsistence

4. CONTAMINANTS

Although persistent organic pollutants (POPs) such as PCBs and DDT are widely banned today, they were used extensively in the middle of the last century. These long-lived pollutants are transported in the air from their source to cooler areas where they condense and are deposit- ed in glacial ice. Until recently, these compounds had remained trapped in the ice, but rapid melting has begun to release them back into the environment [4]. For example, in one Canadian lake, glacial meltwater is the source of 50-97% of the various POPs entering the lake17. At least 10% of this glacial melt is from ice that was deposited between the 1950s and 1970s, as shown by the presence of tritium, a by-product of nuclear bomb tests conducted during this era.

Figure 2: Pesticides used in temperate and tropical areas are transported to the Arctic and deposited in glacial ice.

5. THE ARCTIC

Over recent decades, Arctic glaciers have generally been shrinking, with the exception of Scandinavia and Iceland, where increased precipitation has resulted in a positive balance36. Arctic melting appears to have accelerated in the late 1990s; estimates of combined annual melting rose from 100 sq km per year from 1980-89 to 320 sq km in 1997 and 540 sq km in 199837. Greenland alone contains 12% of the world's ice shown in figure 3.

While portions of the interior are gaining mass, there has been significant thinning and ice loss around the periphery [5]. This loss is not simply due to melting at the edges; entire portions of the Greenland ice sheet appear to be sliding towards the sea. Because this sliding accelerates when surface melting is most intense, it is believed that surface meltwater may be trickling down to the glacial bed and lubricating ice sheet movement [6]. This recent discovery provides a mechanism for rapid response of ice sheets to climate change, a process that was previously believed to require hundreds or thousands of years.

Figure 3: Exit Glacier from camp ground Kenai Fjords National Park Alaska, USA

6. SOUTH AMERICA

The northern Andes contain the largest concentration of glaciers in the tropics, but these glaciers are receding rapidly and losses have accelerated during the 1990s. In Peru, Yanamarey Glacier lost a quarter of its area during the last fifty years25, and Uruashraju and Broggi Glaciers lost 40-50% of their length from 1948-199026. In Ecuador, Antizana Glacier shrank 7-8 times faster during the 1990s than in previous decades. Similarly, Glacier Chacaltaya (Bolivia) lost nearly half of its area and two thirds of its volume during the mid- 1990s alone, and could disappear by 201027. In the sub-tropical wet Andes, the large ice mass- es of the North Patagonia Icefield (Chile) and South Patagonia Icefield (Chile and Argentina) had lost only 4-6% of their 1945 area by the mid 1990s28, but thinning has accelerated recent- ly. Parts of the southern icefield experienced thinning rates from 1995-2000 that were over twice as fast as their average rates during the previous three decades [7]. According to recent research by NASA, the Patagonia icefields of Chile and Argentina are thinning at an accelerating pace and now account for nearly 10 percent of global sea-level change from mountain glaciers.

7. ANTARCTICA

Antarctica is blanketed by ice sheets that contain about 95% of the planets freshwater. Cold temperatures prevent significant surface melting, but recent work shows that bottom melting underneath glaciers at the junction between land and sea is rapid and widespread throughout Antarctica, possibly due to increased ocean temperatures [8]. Warmer seas have also contributed to the rapid thinning and breakup of many large, floating ice shelves. These shelves may buttress and slow the glaciers flowing into them; although there was no change in glacier velocity after the loss of the Wordie Ice Shelf, several major ice streams that nourished the Larsen A shelf are flowing as much as 2-3 times faster towards the sea since its breakup in 199540. At the same time, the interior has experienced an increase in accumulation because more water is being evaporated from warmer seas and falling as snow figure 4. The extent to which these gains com- pensate for ice loss at the edges is unknown. Its glaciers flow across the continent towards the coast where the ice melts or breaks away to produce icebergs.

Figure: 4 Of all the world's glaciers, Antarctic ice will contribute the most significantly to global sea level rise

8. EUROPE

In the past four decades, the majority of glaciers in the Alps have experienced considerable mass losses; this is illustrated by the Hintereisferner (Austria), Gries (Switzerland), and Sarennes (France), each of which lost the equivalent of 14 m ice thickness since the 1960s. Glacier melting has accelerated since 1980, and 10-20% of glacier ice in the Alps was lost in less than two decades18. The discovery of a 5300-year old ice man in a melting glacier in Italy demonstrates that many glaciers are now smaller than they have been for thousands of years [9]. The World Meteorological Organization reports that summer 2003 temperatures, which triggered floods, land slides, and the rapid formation of glacial lakes, were the hottest ever recorded in northern and central Europe; if current trends continue, the European Alps will lose major parts of their glacier coverage within the next few decades [10].

Figure 5: Glaciers and stream Cogne Valley. Grand Paradiso National Park. Italy

9. AFRICA

Tropical glaciers in Africa have decreased in area by 60-70% on average since the early 1900s. The ice fields atop Mt. Kilimanjaro have lost 80% of their area during this century and despite persisting for over 10,000 years, they are likely to disappear by 202019. On Mt. Kenya, 7 of the 18 glaciers present in 1900 had disappeared by 199320, and four glaciers (Lewis, Tyndall, Gregory and Cesar) had lost between 60% and 92% of their area [11]. The remaining glaciers in the Ruwenzori Mountains of Uganda and the Democratic Republic of Congo are also melt- ing rapidly, with area losses during the 20th century ranging from 53% (Speke) to 90% (Moore)21.

Figure 6 : Melting Snows of Kilimanjaro

10. ASIA

The vast majority of all Himalayan glaciers have been retreating and thinning over the past 30 years, with accelerated losses in the last decade. For example, glaciers in the Bhutan Himalayas are now retreating at an average rate of 30-40 m per year [12]. In Central Asia, glaciers are wasting at exceptionally high rates. In the northern Tien Shan (Kazakhstan), glaciers have been collectively losing 2 sq km of ice (0.7% of their total mass) per year since 1955, and Tuyuksu glacier has receded nearly a kilometer since 192310. Glaciers in the Ak-shirak Range (Kyrgyzstan) have lost 23% of their area since 197723, similar to area losses in the northern Tien Shan (29% from 1955-1990) and the Pamirs (16% from 1957-1980). In the Chinese Tien Shan, Urumqihe Glacier lost the equivalent of 4 m ice thickness from 1979-199524, and the Chinese Meteorological Administration pre- dicts that China's northwestern mountains will lose over a quarter of their current glacier cov- erage by 2050. These glaciers supply 15-20% of the water to over 20 million people in the Xinjiang and Qinghai Provinces alone [13]

Figure 7: Himalayan Glacier Lake. Khumbu valley, Annapurna region, Nepal

Figure 7 shown the drinking water is already scarce in parts of Asia where Himalayan glaciers supply freshwater to a third of the world's population [14]. In Bhutan, wooden troughs made from hollowed tree branches carry glacial runoff to communities

11. WATER SHORTAGES

Seventy percent of the world's freshwater is frozen in glaciers, which buffer ecosystems against climate variabili- ty by releasing water during dry seasons or years. In tropical areas, glaciers melt year-round, contributing contin- uously to streamflow and often providing the only source of water for humans and wildlife during dry parts of the year. Freshwater is already a limiting resource for much of the planet, and in the next 30 years population growth is likely to far exceed any potential increases in available water [15].

The Himalayan glaciers that feed seven of the great rivers of Asia (the Ganga, Indus, Brahmaputra, Salween, Mekong, Yangtze and Huang He) and ensure a year-round water supply to 2 billion people are retreating at a startlingly fast rate. In the Ganga, the loss of glacier meltwater would reduce July-September flows by two thirds, causing water shortages for 500 million people and 37% of India's irrigated land [16]. In the northern Tien Shan mountains of Kazakhstan, more than 90% of the region's water supply is used for agriculture and 75-80% of river runoff is derived from glaciers and permafrost, which are melting at accelerated rates10. In the dry Andes, glacial meltwater contributes more to river flow than rainfall, even during the rainy season11. Most large cities in Ecuador, Peru and Bolivia rely on meltwater from rapidly disappearing glaciers for their water supply and hydroelectric power, and many communities are already experiencing shortages and conflicts over use [17].

12. FLOODING

Rapid melting of glaciers can lead to flooding of rivers and to the for- mation of glacial meltwater lakes, which may pose an even more serious threat. Continued melting or calving of ice chunks into lakes can cause catastrophic glacial lake outburst floods. In 1985, such a flood at the Dig Tsho (Langmoche) Lake in Nepal killed several people and destroyed bridges, houses, arable land, and a nearly completed hydropower plant4. A recent UNEP study found that 44 glacial lakes in Nepal and Bhutan are in immediate danger of overflowing as a result of climate change [18]. In Peru, a chunk of glacier ice fell into Lake Palcacocha in 1941, causing a flood that killed 7000 people; recent satellite photos reveal that another chunk of loose ice is poised over this lake, threatening the lives of 100,000 people below7.

Figure 8: Flooding caused by runoff from melting glaciers could have disastrous consequences for people living nearby

13. SOLUTIONS

Worldwide, accelerating glacier loss provides independent and startling evidence that global warming is occurring1. It is now clear that the Earth is warming rapidly due to man-made emissions of carbon dioxide and other heat-trap- ping gases, which blanket the planet and cause temperatures to rise [19]. Climate change is already happening, but we can strive to keep global warming within tolerable limits if we act now.

Based on scenarios of projected damage to ecosystems and human communities, WWF seeks to limit global warm- ing to a maximum of 2°C over pre-industrial levels. Although a warming of 1-2°C will clearly threaten human health, water supplies and vulnerable ecosystems, a warming of at least

1°C appears unavoidable. Warming beyond 2°C is likely to result in rapidly escalating damages, with severe threats to human populations and the loss of unique and irreplaceable ecosystems. It is therefore imperative that emissions of the main heat-trapping gas, carbon diox- ide (CO_2), are significantly reduced, in order to avoid exceeding this 2°C threshold [20].

The majority of CO_2 pollution is released when fossil fuels such as coal, oil and natural gas are burned for trans- portation, heating, or the production of electricity. Coal is particularly damaging, as it produces 70% more CO_2 emissions than natural gas for the same energy output. Electricity generation is the single largest source of man- made CO_2, amounting to 37% of worldwide emissions.

WWF is challenging the electric power sector to become CO_2-free by the middle of this century in industrialized countries, and to make a significant shift towards that goal in developing countries [21]. A number of power companies have already signed on to WWF's vision, but in order to reduce emissions significantly, power utilities, financial institutions, consumers, and policy makers must all play a role:

- Utilities can support meaningful global warming legislation, improve the energy efficiency of power plants, increase their use of renewable energy sources, and halt investment in new coal plants and coal mining.

- Financial institutions can call upon the companies they invest in to disclose their emissions policies, and switch their investments to companies that are striving to be more competitive under future limits on carbon emissions.

- Electricity consumers should opt for "green power" where it is available, demand this choice where it is not, and invest in highly efficient appliances.

- Policy makers must ease the transition to a carbon-free energy industry by passing legislation that creates favorable market conditions, shaping new frameworks for change, and ensuring that the Kyoto Protocol, the world's primary legal tool to combat global warming, enters into force as soon as possible.

14. CONCLUSION

The increase in average temperature will have the direct impact on glaciers and glacial lakes in Hindu Kush-Himalayan region. The glaciers of the Hindu Kush-Himalayan region are retreating and as a result the glacial lakes associated with the glaciers are increasing in number and size to the level of potential glacial lake outburst flood. Many glacial lake outburst flood are recorded in region at least one in 3 to 10 years since 1970s. The glacial lake outburst flood events have trans-boundary effect resulting loss of many lives and property along the downstream. The International Centre for Integrated Mountain Development (ICIMOD) with its partner institutes mapped about 15,000 glaciers, 9000 lakes and 200 potentially dangerous glacial lakes including 21 glacial lake outburst flood events in the Himalayan region except Arunanchal and Azad Jammu & Kashmir (AJK) region. The database of glaciers, glacial lakes, and glacial lake outburst flood in Hindu Kush-Himalayan region serves as the baseline data and information for climate change study, planning for water resource development, to understand and mitigate glacial lake outburst flood associated hazards, thus linking science to policy.

REFERENCES

[1] Fujita, K., Kadota, T., Rana, B., Shrestha, R. B., and Ageta, Y., (2001) 'Shrinkage of Glacier AX010 in Shorong region, Nepal Himalayas in the 1990s'. In Glaciological Research 18, pp 51-54.

[2] Abid, M., et al. 2015. Farmers' Perceptions of Adaptation Strategies to Climate Change and their Determinants: The Case of Punjab Province, Pakistan. Earth System Dynamics. 6(1). pp. 225–243.

[3] Abid, M., et al. 2016. Climate Change Vulnerability, Adaptation and Risk Perceptions at Farm Level in Punjab, Pakistan. Science of the Total Environment. 547. pp. 447–460.

[4] Acemoglu, D., et al. 2012. The Network Origins of Aggregate Fluctuations. Econometrica. 80(5). pp. 1977–2016.

[5] Acemoglu, D., A. Ozdaglar, and N. Tahbaz-Salehi. 2014. Microeconomic Origins of Macroeconomic Tail Risks. NBER Working Paper No. 20865. Washington, DC.

[6] Adams, H., and W. N. Adger. 2013. The Contribution of Ecosystem Services to Place Utility as a Determinant of Migration Decision-Making. Environment Research Letters. 8(1).

[7] Adger, W. N., et al. 2015. Focus on Environmental Risks and Migration: Causes and Consequences. Environmental Research Letters. 10(6). 60201.

[8] Bandara, J. S., and Y. Cai. 2014. The Impact of Climate Change on Food Crop Productivity, Food Prices and Food Security in South Asia. Economic Analysis and Policy. 44(4). pp. 451–465.

[9] Bangladesh Bureau of Statistics. 2015. Population Density and Vulnerability: A Challenge for Sustainable Development of Bangladesh. Dhaka: Bangladesh Bureau of Statistics.

[10] Bardsley, D. K., and G. J. Hugo. 2010. Migration and Climate Change: Examining Thresholds of Change to Guide Effective Adaptation Decision-Making. Population and Environment. 32(2–3). pp. 238–262

[11] Critical policy interventions to fast forward micro irrigation in India Mr Qazi Syed Wamiq Ali and Mr Nathaniel B Dkhar 2019

[12] Ageta, Y., Iwata, S., Yabuki H., Naito, N., Sakai, A., Narama, C. and Karma, (2000) 'Expansion of glacier lakes in recent decades in the Bhutan Himalayas'. In Debris-covered glaciers. IAHS, publ. no. 264, 165 – 175p.

[13] Asahi K., Kadota T., Naito N., and Ageta Y., (2006) 'Variations of small glaciers since the 1970s to 2004 in Khumbu and Shorang regions, eastern Nepal'. In Data Report 4 (2001-2004). Glaciological Expedition in Nepal (GEN) and Cryosphere Research in the Himalaya (CREH). Graduate School of Eviromnmentl Studied, Nagoya University and Department of Hydrology and Meteorology, HMG of Nepal. 109 – 136p.

[14] Bajracharya, S. R.; Mool, P. K.; Shrestha, B. R., (2007) Impact of Climate Change on Himalayan Glaciers and Glacial Lakes: Case Studies on GLOF and Associated Hazards in Nepal and Bhutan. Kathmandu, ICIMOD, 136p.

[15] Bajracharya S. R., and Mool P. K., (2006) Impact of global climate change from 1970s to 2000s on the glaciers and glacial lakes in Tamor Basin, eastern Nepal, ICIMOD.

[16] Bajracharya S. R., and Mool P. K., (2005) 'Growth of hazardous glacial lakes in Nepal' In Proceedings of the JICA Regional seminar on natural disaster mitigation and issues on technology transfer in south and southeast Asia. Sep 30 to 13 October 2004, Kathmandu, TriChandra Campus, Tribhuvan University, 131-148p. Center for Science and Environment, (2002) 'Melting into Oblivion' In Down To Earth, 15 May 2002.

[17] China Daily, 23 September 2004 Dobhal, D.P, Gergan, J.T., Thayyen, R.J. (1999) 'Recession of Dokriani Glacier, Garhwal Himalaya' - An overview. Symp. On snow, ice and glaciers. In A Himalayan Perspective. Geol. Survey India, Abst. Vol. 30-33p.

[18] Eberhard F., (2005) 'Climate Review' Munich Re, Topics Geo, In Annual Review: Natural catastrophes 2005. Knowledge Series, 51p.

[19] A full list of literature cited in this brochure can be found at http://www.panda.org/climate/glaciers

[20] Text by Stacey Combes, Michael L. Prentice, Lara Hansen and Lynn Rosentrater Designed by IPMA Worldwide Press

[21] Website: http://www.panda.org/climate

4

Impacts of Climate Change and Himalayan Melting Glaciers on Human Systems

Laxmi Kumari[*1], Ved Kumar[1,2], Sudhaker Dixit[1], Ajay Singh Yadav[1], P.K. Singh[1], B.R. Singh[2,3]
[1]Department of Humanities and applied Science, School of Management Sciences Lucknow
[2]Center of Vedic Science School of Management Sciences Lucknow
[3]Director General Technical School of Management Sciences Lucknow
[*]e-mail: laxmikumari@smslucknow.ac.in

ABSTRACT

Temperatures in India have risen by 0.7 °C (1.3 °F) between 1901 and 2022, thereby changing the climate in India. In 2022-23 severe heat wave was recorded in Asian country, including India. The temperature reached 51 °C. Climate change makes such heat waves 100 times more likely. The effect of this is falling on the Himalayan glacier, and the glacier is melting. Climate change, with its adverse effects has reached our doorsteps. It is high time that humans act wisely to cope with the changes and take precautions to avoid disasters in the future and save the environment and Himalayan glaciers. Climate change is threatening India's food security with frequent dry spells, heat waves and erratic monsoonal rainfall worsening the farmer's distress. Thus feeding more people more sustainably has become more critical than ever. In this context, India needs to take concrete steps to reduce the impact of global warming. Glaciers have been retreating for over 100 years worldwide, with few exceptions. In international climate observation systems, they are often called unique demonstration objects of climate change. Everybody can see the change and understand that ice melts when it gets warming.

Keywords: Climate Change, Glaciers, Temperature, Greenhouse gases, Human system.

1. INTRODUCTION

Since industrialization and human activities are advancing, the concentration of greenhouse gases in the atmosphere is steadily increasing. As a result of greenhouse gas effect, the world's average surface temperature has increased between 0.3 and 0.6°C over the past hundred years. There is the expectation of a global average temperature increase of 1.4 to 5.8°C in 2100 with the rise of carbon dioxide. The increase in average temperature will directly impact glaciers and glacial lakes in Hindu Kush-Himalayan (HKH) region shown in figure 1. The glaciers of the HKH region are retreating, and as a result the glacial lakes associated with the glaciers are increasing in number and size to the level of potential glacial lake outburst flood. Many GLOFs have been recorded in region at least one in 3 to 10 years since the 1970s. The GLOF events have trans-boundary effect resulting in loss of many lives and property the downstream. The International Centre for Integrated Mountain Development (ICIMOD) with its partner institutes mapped about 15,000 glaciers, 9000 lakes and 200 potentially dangerous glacial lakes including 21 GLOF events in the Himalayan region except for Arunanchal and Azad Jammu & Kashmir (AJK) region [1]. The database of glaciers, glacial lakes, and glacial lake outburst floods in the HKH region serves as the baseline data and information for climate change study, planning for water resource development, to understand and mitigate GLOF associated hazards, thus linking science to policy. However, with the view of catastrophic events of GLOF in the past monitoring, mitigation and awareness of potential GLOF in the region is necessary to reduce the GLOF hazard.

Figure 1: Glacial Shrinking by Increasing Temperature

The U.S. Environmental Protection Agency (EPA) has pointed out that human activities after the Industrial Revolution have released large amounts of carbon dioxide (CO2) and other greenhouse gases into the atmosphere, changing the earth's climate. India is the second most populous country at the tipping point of global warming-induced natural disasters [2].

Atmospheric carbon dioxide levels have increased by over 40%, from 280 ppm in the 18th century to 414 ppm in 2020. While the Industrial revolution has helped reduce poverty, it has also led to a rise in atmospheric carbon dioxide and greenhouse gases. If no proper steps are taken, climate change will take almost 80 million lives in 80 years.

Mountains are amongst the most fragile environments on Earth. They are prosperous repositories of biodiversity, water and providers of ecosystem goods and services on which regional and downstream communities rely. The transport of atmospheric pollutants and climate-altering substances can significantly impact high mountain areas, generally considered "clean" regions. The snow glaciers of the Himalayas, considered the "third pole", one of the largest stores of water on the planet and accelerated melting could have far-reaching effects, such as flooding in the short-term and water shortages in the long-term as the glaciers shrink shown in figure 2.

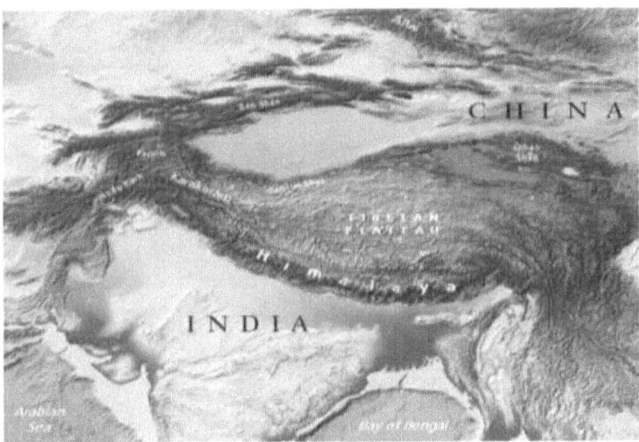

Figure 2: Shrinking Asian Himalayan Glaciers

The data on temperature in Himalayas indicate that warming during the last 3-4 decades has been more than the global average over the last century. Some values show that the Himalayas are warming 5-6 times more than the worldwide average. Mountain systems are seen globally as the prime sufferers of climate change [3]. There is a severe gap in the knowledge of the short and long-term implications of the impact of climate change on water and hazards in the Himalayas, and their downstream river basins. Most studies have excluded the Himalayan region because of its extreme and complex topography and the lack of adequate rain gauge data. There is an urgent need to close the knowledge gap by establishing monitoring schemes for snow, ice and water; downscaling climate models; applying hydrological models to predict water availability; and developing basin-wide scenarios, which also take water demand and socioeconomic development into account. Climate change-induced hazards such as floods, landslides and droughts will impose considerable stresses on the livelihoods of mountain people and downstream populations [4]. Enhancing resilience and promoting adaptation in mountain areas have thus become among the most important priorities of this decade. It is essential to strengthen local knowledge, innovations and practices within social and ecological systems and strengthen the functioning of institutions relevant to adaptation. A common understanding of climate change needs to be developed through regional and local-scale research so that mitigation and adaptation strategies can be identified and implemented.

2. ASIAN COUNTRIES

The vast majority of all Himalayan glaciers have been retreating and thinning over the past 30 years, with accelerated losses in the last decade. For example, glaciers in the Bhutan Himalayas are now retreating at an average rate of 30-40 m per year shown in figure 3. In Central Asia, glaciers are wasting at exceptionally high rates. In the northern Tien Shan (Kazakhstan), glaciers have been collectively losing two sq km of ice (0.7% of their total mass) per year since 1955, and the Tuyuksu glacier has receded nearly a kilometer since 192310.

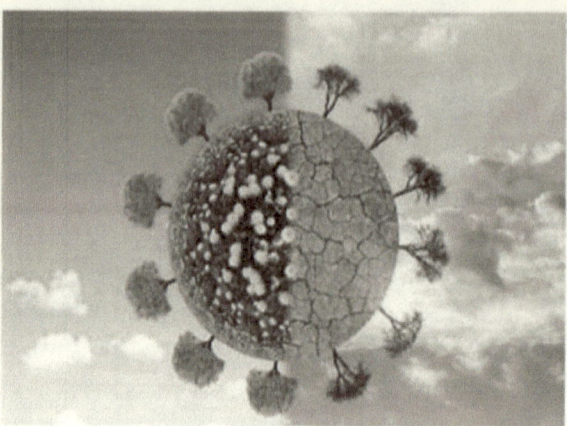

Figure 3: Effect of temperature on Glacier

Glaciers in the Ak-Shirak Range (Kyrgyzstan) have lost 23% of their area since 1977, similar to area losses in the northern Tien Shan (29% from 1955-1990) and the Pamirs (16% from 1957-1980). In the Chinese Tien Shan, Urumqi he Glacier lost the equivalent of 4 m ice thickness from 1979-1995. The Chinese Meteorological Administration predicts that China's northwestern mountains will lose over a quarter of their current glacier coverage by 2050. These glaciers supply 15-20% of the water to over 20 million people in the Xinjiang and Qinghai provinces alone [6].

3. NEED ECO-FRIENDLY AGRICULTURE IN THE WORLD

As the global population rapidly approaches 8 billion, more pressure than ever is on the agriculture industry to increase production. Climate change is threatening India's food security with frequent dry spells, heat waves and erratic monsoonal rainfall worsening the farmer's distress and thus, feeding more people more sustainably has become more critical than ever. As per the Report 'Our World in Data', food production contributes more than a quarter (26%) of global greenhouse gas emissions; Agriculture is also the primary driver of biodiversity loss, threatening 86% of species at risk of extinction, according to the UN. This impact is likely to increase as the world's population grows [7].

4. EMISSIONS IN INDIA FROM AGRICULTURE

India's annual GHG emissions from agriculture and livestock stand at 481 megaton's of CO_2 equivalent, of which 42% come from crop production and 58% is emitted from livestock. Cattle production is the highest source of emissions, followed by rice the highest emitter among the crops with 52% of all crop-related emissions. Rice is a water-loving plant and unlike other crops, it requires standing water in the field. Such conditions are ideal for methane-producing bacteria to produce methane, a gas with a global warming potential that is 56 times higher than that of carbon dioxide over 20 years. It is also released as a by-product of bacterial fermentation in the digestive tract of cattle, buffaloes and other ruminants and is belched out by the animals [8]. Thus it becomes important that India try to reduce its carbon footprint as much as possible, more in its farming sector. Creating sustainable and climate-resilient agricultural systems has been highlighted as part of India's plan to meet its ambitious pledge to the United Nations Framework Convention on Climate Change international treaty to reduce the emission intensity of its GDP by up to 35% by 2030, compared to 2005 levels. The consistent picture that emerges is net ice loss in most parts of the Himalayas emissions in India from agriculture figure 4. Measurements suggest that the rate of loss has increased since about 2020.

Global Warming is melting glaciers in every region of the world, putting millions of people at risk from floods, droughts and lack of drinking water. Glaciers are ancient rivers of compressed snow that creep through the landscape, shaping the planet's surface. They are the Earth's largest freshwater reservoir, collectively covering an area the size of South America. Glaciers have been retreating worldwide since the end of the Little Ice Age (around 1850), but in recent decades glaciers have begun melting at rates that cannot be explained by historical trends1.

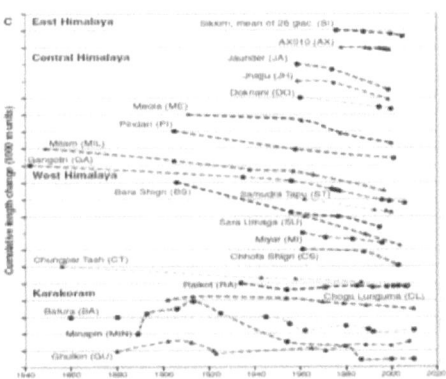

Figure 4 : Emissions in India from Agriculture

Projected climate change over the next century will further affect the rate at which glaciers melt. Average global temperatures are expected to rise 1.4-5.8°C by the end of the 21st century [9]. Simulations project that a 4°C rise in temperature would eliminate nearly all of the world's glaciers (the melt-down of the Greenland ice sheets could be triggered at a temperature increase of 2 to 3°C). Even in the least damaging scenario – a 1°C rise and an increase in rain and snow-glaciers will continue to lose volume over the coming century. Although only a tiny fraction of the planet's permanent ice is stored outside of Greenland and Antarctica, these glaciers are significant because they respond rapidly to climate change and their loss directly affects human populations and ecosystems. Continued, widespread melting of glaciers during the coming century will lead to floods, water shortages for millions of people, and sea level rise, threatening and destroying coastal communities and habitats.

5. HABITAT LOSS

While many species are likely to be affected by changes in stream flow and sea level associated with glacier melting, animals that dwell on or near glaciers may be pushed towards extinction by the disappearance of their icy habitats. Far from being barren expanses of ice, glaciers are home to some of the unique organisms and ecosystems.

Figure 5: Habitat loss of glacier animals

For example, the tiny ice worm spends its entire life on ice, roaming over glaciers at night, feeding on glacial algae, and occasionally being snatched up by a hungry snow bunting. The physiological adaptation that allows these worms to survive at 0°C remains unknown [10]. Because these worms disintegrate at temperatures over 5°C, their secret may be lost as temperatures rise and their glacial habitat melts away shown figure 5. Climate change has already led to the loss of an entire ecosystem on the crumbling ice shelves of the Arctic. Between 2000 and 2002, the Ward Hunt Ice Shelf off of Ellesmere Island in Canada broke in two, draining much of the water from overlying Disraeli Fjord, the largest remaining epi shelf (ice shelf-bounded) lake in the Northern Hemisphere. This 3000-year-old lake supported a rare ecosystem where microscopic marine organisms near the bottom of the lake lived in harmony with their freshwater brethren in the brackish surface waters. By 2002, 96% of this unique low-salinity habitat had been lost. Even animal not living directly on glaciers can be severely affected by their disappearance. Kittlitz's murrelet, for example, is a small diving seabird that forages for food almost exclusively in areas where glacial melt water enters the ocean.

Figure 6: Habitat loss of glacier animals

These birds are already in serious trouble; their global population (mostly in Alaska) is thought to have plummeted from several hundred thousand in 1972 to less than 20,000 in the early 1990s. Several conservation groups have filed a petition to declare Kittlitz's murrelet an **endangered species**, citing climate change and the loss of critical glacier-associated habitat as one of the primary reasons for the species' decline shown in figure 6. Even farther away from the melting glaciers coral reefs will be affected by rising sea levels. Corals require light for photosynthesis to survive. The depth at which corals can live is limited by how far light can penetrate the water. When light diminishes as the sea level rises, corals living at this light-limiting depth will be lost. Coral reefs at other depths will also see reduced growth rates as light quality changes from rising sea levels. One simulation showed that coral reefs in the Caribbean are not expected to be able to keep up with sea level rise [11]. This has consequences not only for the corals and marine life but for the human communities that rely on these reefs for subsistence.

6. CONTAMINANTS

Although persistent organic pollutants (POPs) such as PCBs and DDT are widely banned today, they were used extensively in the middle of the last century. These long-lived pollutants are transported from their source to more excellent areas where they condense and are deposited in glacial ice. Until recently, these compounds had remained trapped in the ice, but rapid melting has begun to release them back into the environment. For example, in one Canadian lake, glacial melt water is the source of 50-97% of the various POPs entering the lake17. At least 10% of this glacial melt is from ice deposited between the 1950s and 1970s, as shown by the presence of tritium, a by-product of nuclear bomb tests conducted during this era [12].

7. EUROPE

In the past four decades, the majority of glaciers in the Alps have experienced considerable mass losses; this is illustrated by the Hintereisferner (Austria), Gries (Switzerland), and Sarennes (France), each of which lost the equivalent of 14 m ice thickness since the 1960s.

Figure 7: Effect of Temperature on Glacier Melting

Glacier melting has accelerated since 1980, and 10-20% of glacier ice in the Alps was lost in less than two decades1shown in figure 7. The discovery of a 5300-year-old iceman in a melting glacier in Italy demonstrates that many glaciers are now smaller than they have been for thousands of years. The World Meteorological Organization reports that summer 2003 temperatures, which triggered floods, landslides, and the rapid formation of glacial lakes, were the hottest ever recorded in northern and central Europe; if current trends continue, the European Alps will lose major parts of their glacier coverage within the next few decades [13].

8. ANTARCTICA

Antarctica is blanketed by ice sheets that contain about 95% of the planets freshwater. Cold temperatures prevent significant surface melting. However, recent work shows that bottom melting underneath glaciers at the junction between land and sea is rapid and widespread throughout Antarctica, possibly due to increased ocean temperatures [14]. Warmer seas have also contributed to the rapid thinning and breakup of many large, floating ice shelves. These shelves may buttress and slow the glaciers flowing into them. However, there was no change in glacier velocity after the loss of the Wordie Ice Shelf, several major ice streams that nourished the Larsen shelf are flowing as much as 2-3 times faster towards the oceans since its breakup in 199540. At the same time, the interior has experienced an increase in accumulation because more water is being evaporated from warmer seas and falling as snow. The extent to which these gains compensate for ice loss at the edges is unknown.

9. WATER SHORTAGES

Seventy percent of the worlds freshwater are frozen in glaciers, which buffer ecosystems against climate variability by releasing water during dry seasons or years. In tropical areas, glaciers melt year-round, contributing continuously to stream flow and often providing the only water source for humans and wildlife during dry parts of the year.

Figure 8: Effect of temperature on water shortage

Freshwater is already a limiting resource for much of the planet, and in the next 30 years, population growth is likely to exceed any potential increases in available water far in figure 8. The Himalayan glaciers that feed seven of the great rivers of Asia (the Ganga, Indus, Brahmaputra, Salween, Mekong, Yangtze and Huang He) ensure year-round water supply to 2 billion people are retreating at a startlingly fast rate. In the Ganga, the loss of glacier melt water would reduce July-September flows by two-thirds, causing water shortages for 500 million people and 37% of India's irrigated land. In the northern Tien Shan Mountains of Kazakhstan, more than 90% of the region's water supply is used for agriculture and 75-80% of river runoff is derived from glaciers and permafrost, melting at accelerated rates. In the dry Andes, glacial melt water contributes more to river flow than rainfall, even during the rainy season [15]. Most large cities in Ecuador, Peru and Bolivia rely on melt water from rapidly disappearing glaciers for their water supply and hydroelectric power. Many communities are already experiencing shortages and conflicts over use.

10. FLOODING

Rapid melting of glaciers can lead to the flooding of rivers and to the formation of glacial melt water lakes, which may pose an even more severe threat. Continued melting or calving of ice chunks into lakes can cause catastrophic glacial lake outburst floods. In 1985, a flood at the Dig Tsho (Langmoche) Lake in Nepal killed several people and destroyed bridges, houses, arable land, and a nearly completed hydropower plant [16]. A recent UNEP study found that 44 glacial lakes in Nepal and Bhutan are in immediate danger of overflowing due to climate change 5, 6. In Peru, a chunk of glacier ice fell into Lake Palcacocha in 1941, causing a flood that killed 7000 people; recent satellite photos reveal that another piece of loose ice is poised over this lake, threatening the lives of 100,000 people below.

11. SEA LEVEL RISE

The average global sea level rose by 1-2 mm per year during the 1900s and is projected to continue growing, with an estimated contribution of 0.2-0.4 mm per year from melting glaciers. The effect of glaciers may be underestimated; however, recent studies suggest that accelerated melting in Alaska and the Patagonia Ice fields since the mid-1990s has increased the combined contribution of just these two areas to 0.375 mm per year. Sea-level rise will affect coastal regions worldwide, causing flooding, erosion, and saltwater intrusion into aquifers and freshwater habitats. Even the modest sea-level rise during the 20th century led to decline and the loss of 100 sq km of wetlands yearly in the U.S. Mississippi River Delta15. In Trinidad and Tobago, as in many low-lying islands, beaches are retreating several meters per year and salinity levels have begun to rise in coastal aquifers. Small Pacific islands such as Tonga, the Marshall Islands and the Federated States of Micronesia are particularly vulnerable [17]. They could lose significant portions of their land area to rising seas and storm surges. A global sea level rise of 1 m would inundate 80% of the Maldives, displace 24 million people in Bangladesh, India and Indonesia, and eliminate the Sundarbans, the world's largest mangrove forest and home to the endangered Royal Bengal Tiger and hundreds of other species.

12. CLIMATE CHANGE, GLOBAL WARMING AND THEIR EFFECTS

Learn about Global warming's root cause of climate change, the role of the greenhouse effect and its impact on the environment.

13. CLIMATE CHANGE

The climate is a long-term process compared to the weather. Our world has undergone several changes over billions of years by many naturally occurring forces. The various natural forces are the ice age, variation in the sun's intensity, volcanic eruptions, naturally occurring greenhouse gas concentrations, glaciations etc [18]. The change which affects human activity directly or indirectly and modifies the configuration of the global atmosphere is called climate change. Over an approximate period, it also brings variability in the natural climate.

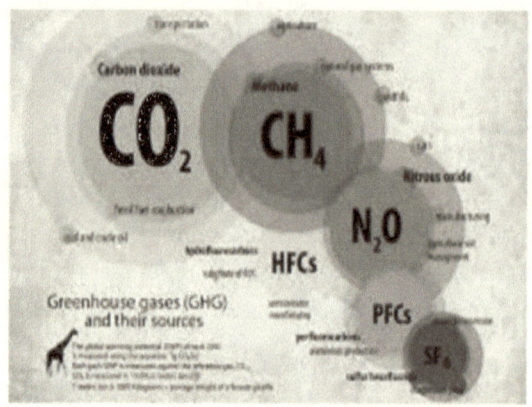

Figure 8: Greenhouse effect and its impact on the environment

Climate change is a long-time change in weather. This change won't happen in a day. Climate change harms warming trends. It also affects the pattern of rainfall, temperature, wind and snowfall. Human activities such as deforestation and overconsumption of fossil fuels are also responsible for climate change [19]. An increase in average atmospheric temperature on the earth's surface is called global warming. Climate patterns get affected due to the effects of global warming in the troposphere. Global warming is mainly due to global activity of emission of greenhouse gases by industries, vehicles, and many others.

14. GREENHOUSE EFFECT

One of the consequences of the greenhouse effect is a warmer temperature in the earth's lower atmosphere which makes it inappropriate for survival. There are many causes of the greenhouse effect, like deforestation, greenhouse gases and many more [20].

Table-1: Causes of greenhouse gases

Greenhouse Gases	Description
Water Vapour	1. Major contributor among the greenhouse gases 2. But it won't be present in the air for too long 3. Present in the atmosphere
Carbon Dioxide	4. Natural compound present in the atmosphere 5. Also created by human activities like burning of fossils, deforestation, etc 6. It is considered the most common greenhouse gas
Nitrous Oxide	1. Natural compound present in the atmosphere 2. Combustion of fossil fuels, agricultural activities and treatment of wastewater etc. releases nitrous oxide
Fluorinated gases	1. Hydro fluorocarbons (HFC) 2. Per fluorocarbons (PFC) released from semiconductors and aluminum processing 3. Sculpture Hexafluoride released during magnesium processing 4. It can cause severe effects on global warming, it can be removed only by the more amount of sunlight that is available in the upper atmosphere
Methane	1. Natural compound present in the atmosphere 2. Emitted by livestock, released by wetland and from natural gas leakage
Black carbon	1. The incomplete combustion of organic matter results in the formation of soot and these particles are made of different varieties of pure carbon 2. It has a high tendency to absorb sunlight and is present in the atmosphere for a longer period
Brown carbon	1. The major source of brown carbon is biomass combustion, and it is considered soot and dust particles obtained from incomplete combustion

15. IMPACT OF CLIMATIC CHANGE

15.1 Agricultural and Food Security

Increase in greenhouse gases affects the cropping pattern from region to region. For example, in the temperate region, moderate warming (rise of 1 to 3°C in mean temperature) can benefit crop production. But in dry tropical regions, the rise of 1 to 2 °C can have a negative influence on the major cereal crops. Increase in temperature of more than 3 degrees can severely affect crop production.

15.2 Sea Levels, Oceans and Coastal Areas

Global temperatures have risen about 1 degree Celsius (1.8 degrees Fahrenheit) over the last 100 years, with a sea-level response of 160 to 210 mm (about half of that amount occurring since 1993), or about 6 to 8 inches. Many effects have been observed in recent years, including an increase in sea-surface temperature, significant evaporation, glacier melting, and noticeable change in the marine food web. By the year 2100, the sea level will increase by 9 to 88 cm. It is due to the inundation of fresh water from the melting of glaciers and the change in temperature that causes seawater warming. Coastal areas and smaller islands are more prone to the effects of increased sea levels as these areas are more prone to floods and erosion. Saltwater intrusion is one of the major effects of an increase in extreme sea-level events like a tsunami. High tides and heavy storms are also results of sea level rising [21].

15.3 Water Stress and Water Security

Warming has created a great impact on glaciers and snow cover in both hemispheres. This prognosis to speed up throughout the 21st century. It has led to the depletion of water resources and hence the potential of hydropower. It also has changed the seasonal flow of water due to the supply of freshwater from glaciers and major mountain ranges (Hindu-Kush, Himalayas, Andes) By the year 2050, the availability of fresh water in the south, central, east and southeast Asia will also decline. An increase in atmospheric temperature will scramble the hydrologic cycle [22]. It also affects the rainfall rate. During the 20th century, an increase in floods along river basins was noticed. The frequent flooding causes problems to society, water quality, and infrastructure. Damage to public infrastructure impacts a much larger percentage of the population than those whose homes or businesses were directly flooded. Flood damage to roads, rail networks, and critical transportation hubs like maritime ports, in particular, can have a big impact on regional and national economies. The chemical, physical and biological properties of lakes and rivers are affected due to warmer temperatures, which negatively impact freshwater species [23].

15.4 Biological Diversity

Species in the tropical area are at greater risk, according to the World Wildlife Fund (WWF) as the temperature increases, the species starts to migrate to regions of lower temperature. So, the change in temperature forces the species to migrate. According to WWF, one-fifth of the endangered species could be facing a catastrophic loss of biodiversity shown in figure 9. The marine ecology will be severely impacted [24]. This may result in changes to the ocean's circulation, an increase in the ocean's acid range, and a rise in carbon dioxide levels this level of change affects corals and shell-forming animals.

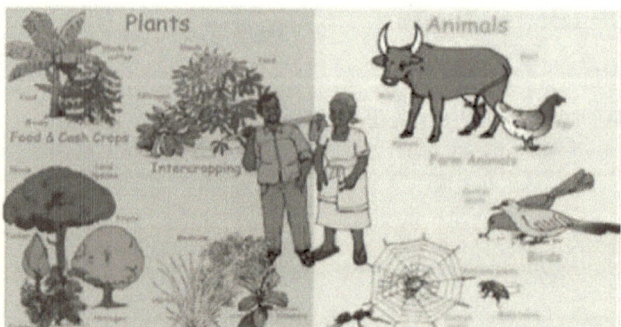

Figure 9: Biological diversity

A temperature rise will also increase the air pollutants in the atmosphere. So many diseases may occur due to air pollutants. Climate change adversely affects human health. It also makes the earth's temperature unfavorable for humans. Increased heat waves and weather events also increase casualty. Climate change is also responsible for the spreading of disease. These diseases spread from one geological region to another. Diversifying from existing cropping systems, predominated by rice and wheat in many unsustainable landscapes, to more nutritious and environment-friendly crops is being suggested to address challenges of climate change and malnutrition. Agro forestry, for example, brings synergies between trees and crops or forages such as trees on field bunds, inline agro forestry and high-density fruit orchards to help diversify existing farming systems and achieve medium to long term sustainability [25]. Diversification to crops like pulses, oilseeds, vegetables and fruits, adapted to specific agro-ecologies, must also be planned, and implemented by the states with suitable incentives to farmers during the changeover.

16. CONCLUSION

The effect of this is falling on the Himalayan glacier, and the glacier is melting. Climate change, with its adverse effects has reached our doorsteps. It is high time that humans act wisely to cope with the changes and take precautions to avoid disasters in the future and save the environment and Himalayan glaciers. Climate change is threatening India's food security with frequent dry spells, heat waves and erratic monsoonal rainfall worsening the farmer's distress. Thus feeding more people more sustainably has become more critical than ever. In this context, India needs to take concrete steps to reduce the impact of global warming. Climate change disproportionately affects the poor and the smallholders, who earn their livelihoods from agriculture. Climate-friendly agriculture offers new income sources and is more sustainable India's carbon emissions could drop by 45-62 million tonnes annually. Nature-positive and regenerative agriculture practices mentioned above hold the potential to reduce GHG emissions. Post-cop26, India's ambitious commitments should reflect in its tangible and concrete actions. Climate change, with its adverse effects, has reached our doorsteps. It is high time that humans act wisely to be able to cope with the changes and take precautions to avoid disasters in future and save the environment.

REFERENCES

[1]   Ageta, Y., Iwata, S., Yabuki H., Naito, N., Sakai, A., Narama, C. and Karma, (2000) 'Expansion of glacier lakes in recent decades in the Bhutan Himalayas'. In Debris-covered glaciers. IAHS, publ. no. 264, 165 – 175p.

[2]   Asahi K., Kadota T., Naito N., and Ageta Y., (2006) 'Variations of small glaciers since the 1970s to 2004 in Khumbu and Shorang regions, eastern Nepal'. In Data Report 4 (2001-2004). Glaciological Expedition in Nepal (GEN) and Cryosphere Research in the Himalaya (CREH). Graduate School of Eviromnmentl Studied, Nagoya University and Department of Hydrology and Meteorology, HMG of Nepal. 109 – 136p.

[3]   Bajracharya, S. R.; Mool, P. K.; Shrestha, B. R., (2007) Impact of Climate Change on Himalayan Glaciers and Glacial Lakes: Case Studies on GLOF and Associated Hazards in Nepal and Bhutan. Kathmandu, ICIMOD, 136p.

[4]   Bajracharya S. R., and Mool P. K., (2006) Impact of global climate change from 1970s to 2000s on the glaciers and glacial lakes in Tamor Basin, eastern Nepal, ICIMOD.

[5]   Bajracharya S. R., and Mool P. K., (2005) 'Growth of hazardous glacial lakes in Nepal' In Proceedings of the JICA Regional seminar on natural disaster mitigation and issues on technology transfer in south and southeast Asia. Sep 30 to 13 October 2004, Kathmandu,

TriChandra Campus, Tribhuvan University, 131-148p. Center for Science and Environment, (2002) 'Melting into Oblivion' In Down To Earth, 15 May 2002.

[6]   China Daily, 23 September 2004 Dobhal, D.P, Gergan, J.T., Thayyen, R.J. (1999) 'Recession of Dokriani Glacier, Garhwal Himalaya' - An overview. Symp. On snow, ice and glaciers. In A Himalayan Perspective. Geol. Survey India, Abst. Vol. 30-33p.

[7]   Eberhard F., (2005) 'Climate Review' Munich Re, Topics Geo, In Annual Review: Natural catastrophes 2005. Knowledge Series, 51p.

[8]   Fujita, K., Kadota, T., Rana, B., Shrestha, R. B., and Ageta, Y., (2001) 'Shrinkage of Glacier AX010 in Shorong region, Nepal Himalayas in the 1990s'. In Glaciological Research 18, pp 51-54.

[9]   Abid, M., et al. 2015. Farmers' Perceptions of Adaptation Strategies to Climate Change and their Determinants: The Case of Punjab Province, Pakistan. Earth System Dynamics. 6(1). pp. 225–243.

[10]   Abid, M., et al. 2016. Climate Change Vulnerability, Adaptation and Risk Perceptions at Farm Level in Punjab, Pakistan. Science of the Total Environment. 547. pp. 447–460.

[11]   Acemoglu, D., et al. 2012. The Network Origins of Aggregate Fluctuations. Econometrica. 80(5). pp. 1977–2016.

[12]   Acemoglu, D., A. Ozdaglar, and N. Tahbaz-Salehi. 2014. Microeconomic Origins of Macroeconomic Tail Risks. NBER Working Paper No. 20865. Washington, DC.

[13]   Adams, H., and W. N. Adger. 2013. The Contribution of Ecosystem Services to Place Utility as a Determinant of Migration Decision-Making. Environment Research Letters. 8(1).

[14]   Adger, W. N., et al. 2015. Focus on Environmental Risks and Migration: Causes and Consequences. Environmental Research Letters. 10(6). 60201.

[15]   Bandara, J. S., and Y. Cai. 2014. The Impact of Climate Change on Food Crop Productivity, Food Prices and Food Security in South Asia. Economic Analysis and Policy. 44(4). pp. 451–465.

[16]   Bangladesh Bureau of Statistics. 2015. Population Density and Vulnerability: A Challenge for Sustainable Development of Bangladesh. Dhaka: Bangladesh Bureau of Statistics.

[17]   Bardsley, D. K., and G. J. Hugo. 2010. Migration and Climate Change: Examining Thresholds of Change to Guide Effective Adaptation Decision-Making. Population and Environment. 32(2–3). pp. 238–262

[18]   Critical policy interventions to fast forward micro irrigation in India Mr Qazi Syed Wamiq Ali and Mr Nathaniel B Dkhar 2019

[19]   Drought Proofing India: Key Learnings from Bundelkhand Drought Mitigation Package Dr J P Mishra and Dr Shresth Tayal 2018

[20]   Aligning India's water resource policies with the SDGs Dr Girija K Bharat and Mr Nathaniel B Dkhar 2018

[21]   Sustainable Urban Development: Necessity of Integrating WaterEnergy-Food Dimensions in Developmental Policies Dr Shresth Tayal and Ms Swati Singh 2018

[22]   Water Neutral Electricity Production in India: Avoiding the Unmanageable Dr Shresth Tayal and Ms Sonia Grover 2016

[23]   Faecal Sludge Management in Urban India: Policies, Practices and Possibilities Dr S K Sarkar and Dr Girija Bharat 2016

[24]   Mr S Vijay Kumar and Dr Girija K Bharat Perspectives on a Water Resource Policy for India TERI 2014

[25]   Enhancing water-use efficiency of thermal power plants in India: need for mandatory water audits Mr Anshuman 2012

5

Role of Renewable Energy in Climate Change on Human Systems

Ved Kumar[1*], Anod Kumar Singh[2], Pramod Kumar Singh[3],
Sudhaker Dixit[4], Ajai Singh Yadav[5] and Dharmendra Singh[6]
[1-]Department of Humanities & Applied Sciences,
School of Management Sciences, Lucknow, U.P., India.
*e-mail: vedchem21@gmail.com

ABSTRACT

Biodiesel production has received considerable attention in the recent past as biodegradable and non-polluting fuel. Substituting petro-diesel with biodiesel may reduce air emission, increase the domestic supply of fuel, and create new market for agriculture. An alternative approach would be one focus on multi-purpose; short-duration annual crops that can either simultaneously yield fuel along with food/fodder or can be cultivated in rotation with food crops, so that there are a lot of opportunities for small formers. The central policy of biodiesel concerns for creation and protection of environment. The economic benefits include support to the agriculture sector tremendous employment opportunities in plantation and processing. Biodiesel fuel plays an important role for the replacement of petro-diesel to eco-friendly fuel. Various studies showed that pollutants like CO, CO_2, SO_X, HC, PAH, PM etc can be reduced by using blended and pure biodiesel. NO_X emissions are increased by using biodiesel. Biodiesel is an environment friendly biofuel since it provide a means to recycle of CO_2; biodiesel does not contribute global warming. Biodiesel is produced from various plant oils like Jatropha oil, Cottonseed oil, Pongania oil, Palm oils, Rapeseed oil, Castor oil and sorghum oil converted to biodiesel through the process of transesterification. This study is initiated to investigate the potential of jatropha oil as source of biodiesel. This review paper describes the production of biofuel, its properties, agricultural benefits, eco-friendly, recycle of CO_2 and reduce greenhouse gases future potential of biodiesel.
Keywords: *Renewable Energy, Agriculture benefit, Sorghum, Jatropha, Transesterification, Eco-friendly.*

1. INTRODUCTION

Rising petroleum prices, increasing threat to environment from exhaust emission, global warming and threat of supply instabilities has led to a growing concern for it thought the world, more so in the petroleum importing countries like India. Petroleum oil reserve is limited but the oil consumption rate is increasing at an alarming rate. Mechanized agriculture food production systems depend heavily on liquid fuels particularly diesel fuel. The role of agriculture as a source of energy resources is gaining in importance (1). Therefore, agricultural scientists are more interested in developing some biomass-based fuels as alternate fuels for diesel engines. Some of such tried include plant oils (2-3), bio-gas (4), producer gas (5), alcohol (6) and many starting materials such as soybean oil (7-8), sunflower oil (9-10), cotton seed oil (11), rapeseed oil (12), palm oil (13-14) and jatropha oil (15) have been evaluated for preparation of biodiesel by alkaline catalytic rout. An alternative approach would be one which focuses on multipurpose short-duration annual crops that can either simultaneously yield fuel along with food and/or fodder or can be cultivated in rotation with food crops, so that even small private farmers can benefit. Some crops that already commercially well know and can be scaled up produce for bio-energy is also discussed.

Energy crops constitute significant potential for meeting the future energy need worldwide. In addition, agricultural lands offer an alternative to the agriculture which is referred to as energy farming. The studies on energy crops in biodiesel production show that they are quite an economical and environmentally beneficial way of sustainable energy production.

An alternative approach would be one focus on multi-purpose; short-duration annual crops that can either simultaneously yield fuel along with food/fodder or can be cultivated in rotation with food crops, so that there are a lot of opportunities for small formers. The central policy of biodiesel concerns for creation and protection of environment. The economic benefits include support to the agriculture sector tremendous employment opportunities in plantation and processing. Biodiesel fuel plays an important role for the replacement of petro-diesel to eco-friendly fuel. Various studies showed that pollutants like CO, CO_2, SO_x, HC, PAH, PM etc can be reduced by using blended and pure biodiesel. NO_x emissions are increased by using biodiesel. Biodiesel is an environment friendly biofuel since it provide a means to recycle of CO_2; biodiesel does not contribute global warming. Biodiesel is produced from various plant oils like Jatropha oil, Cottonseed oil, Pongania oil, Palm oils, Rapeseed oil, Castor oil converted to biodiesel through the process of transesterification. B5 to B100 has been shown to reduce PM, HC, CO, NO_x, SO_2, CO_2 emission percentage [5] shows in figure 1.

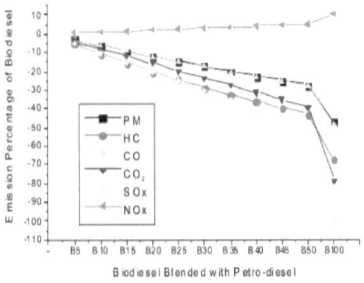

Figure 1: Used biodiesel blend to reduce PM, HC, CO, NOx, SO_x, CO_2 emission percentage

1.1 Future of renewable energy in India

India, faced with twin challenges on energy and environmental front, has no option but to work towards increasing the role of renewable in the future energy systems. Renewable energy technologies vary widely in their technological maturity and commercial status.

In India, renewable energy is at the take-off stage and businesses, industry, government and customers have a large number of issues to address before these technologies could make a real penetration. India with large renewable energy resources (solar PV, wind, solar heating, small hydro and biomass) is to set to have large-scale development and deployment of renewable energy projects [4]. . Renewable energy development is considered in India to be of great importance from the point of view of long term energy supply security, environmental benefits and climate change mitigation. The Integrated Energy Policy report has recognized the need to maximally develop domestic supply options as well as the need to diversify energy sources. MNRE has included in its mission: energy security; increase in the share of clean power; energy availability and access; energy affordability; and energy equity shown in figure 2 [5].

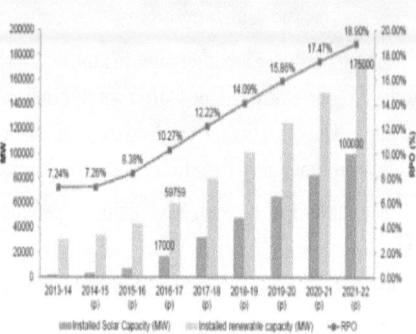

Figure 2: Renewable Energy future in India

1.2 Environment impact of Renewable energy

Renewable energy sources have a large potential to contribute to the sustainable development by providing them with a wide variety of socioeconomic and environmental benefits. Renewable sources of energy are currently unevenly and insufficiently exploited. Although many of them are abundantly available, and the real economic potential considerable, renewable sources of energy make a disappointingly small contribution. There are environmental benefits from renewables other than reduction of greenhouse gas and other air emissions.

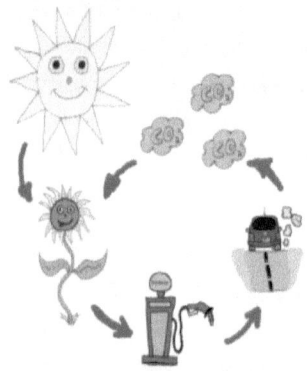

Figure 3: Biofuels Closed Carbon Cycle

Every type of energy utilization for electricity generation has environmental consequences. The main consequences of burning fossil fuels and of nuclear power are well-known. Renewable energy sources (wind, solar, biomass, hydroelectric, geothermal, etc.) are generally thought of as harmless, but this doesn't mean they have no environmental consequences at all. Most of them have a significant aesthetic impact and require large areas of land. Some also have a significant impact on the eco-system shown in fig.3.

2. Sorghum (jowar)

Energy crops constitute significant potential for meeting the future energy need worldwide. In addition, agricultural lands offer an alternative to the agriculture which is referred to as energy farming. The studies on energy crops in biodiesel production show that they are quite an economical and environmentally beneficial way of sustainable energy production.

Sorghum (sorghum bicolor) belongs to Poaceae family, shown in figure 4 and is one of the major varieties of sorghum that has high sugar content. It is similar to sugar cane with sugar rich stalk. It has various uses and is a key of bio-energy crop. Sorghum is considered as a new industrial energy crop as it is an alternative source of energy. It has the potential to be an important alternative feedstock

Figure: 4 Sorghum plant (Jowar)

for the production of biofuel. Sorghum (Sorghum bicolor), popularly called as jowar, Jondhri, Jundi, Chari, is most extensively grown creal grain in country. The crop is environmentally-friendly as it is water-efficient, requires little or no fertilizer or pesticides and biodegradable (7).

2.1 Cultivation of Sorghum (Jowar)

The sorghum grown in two major seasons, viz kharif and rabi. The decline in area is mostly in kharif and at present, area of both kharif and rabi is more or less equal. As far as the productivity is concerned, the kharif crop yield higher when compared with rabi crop (8) .This crop is ideally suited for semi-arid agro-climatic regions of the country and, it gives reasonable good yield with minimal requirement of irrigation and fertilizers (9). Sorghum is an economical crop for production of biofuel, because of its dual nature of sorghum; they produce both grain and sugar rich stalks. Figure 5 show the production of sorghum grain in India (10). Its cultivation cost is also lower than that of sugarcane.

Sorghum crops are genus comparing many species growing in tropical and subtropical countries; eight species are reported to occur in India. Sorghum grain is crop plant, which grown in several parts of India. The food, feed and fodder needs of farmers will not be affected, as the oil extraction Maintaining the Integrity of the Specifications.

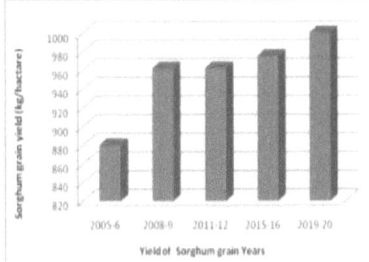

Figure 5: Production of sorghum (jowar) grain in India

3. JATROPHA CURCUS

Jatropha curcas belonging to family Euphorbiaceae, a perennial plant produces seed with 31-37 percent of oil, which can be combusted as fuel without being refined. This fuel known as the process of transesterification can produce bio-diesel. Jatropha plant is grown in marginal and poor soil with minimum cultural practices or in wastelands with low fertility, rockiness, and shallowness of soil. The genus Jatropha has 476 species and distributed throughout the world. Among them 12 species are recorded in India. The species Jatropha curcas is a promising one with economic seed yield and oil recovery.

Figure 6: Jatropha plats , fruits and seeds

Normally the plant is 3-5 meter in height but upto a height of 8 meter has been found under favorable conditions. The fruits are 2.5 centimeter long, black and 2 to 3 halved. Fruits mature by September-October when the capsule changes from green to yellow in colour. It has nearly 420 fruits and 1580 seeds per kg respectively shown in figure 6. Besides higher cetane number, oil reduces emission of carbon monoxide by 44 percent, sulphates by 100 percent and ozone forming potential by less than 50 percent. Through Jatropha cultivation, not only bio-diesel can be obtained but also a tremendous opportunity will be there for employment generation in an agricultural country like India.

Table-1: Fatty acid compositions of crude Jatropha and sorghum oil.

Fatty acid	Formula	Structure	Sorghum %	Jatropha
Palmitic	$C_{16}H_{32}O_2$	16:0	10-14	1.2-7
Stearic	$C_{18}H_{36}O_2$	18:0	3.0-6.0	5.0-9.7
Oleic	$C_{18}H_{34}O_2$	18:1	3.0-47	37-63
Linoleic	$C_{18}H_{32}O_2$	18:2	40-55	19-41
Lenolenic	$C_{18}H_{30}O_2$	-	---	--
Myristic	$C_{14}H_{28}O_2$	14: 0		0.5-1.40

3.1 Fatty acid composition of sorghum and Jatropha oil

The seed kernel of sorghum is comprised of 30-50% oil (10).The fatty composition of sorghum oil consist of palmitic acid 10-14%, stearic acid 3-6%, oleic acid 3-47%, lenoliec acid 40-55% and lenolenic acid 0-1%,[10]. Fatty acid composition of lipids from sorghum oil (19) has given in Table 1, while composition of the oil is similar to other oil, which are edible and non-edible oil purpose. Thus it is good choice as the starting oil for the production of biodiesel.

4. MATERIAL AND METHODS

4.1 Chemicals

Sorghum oil and Jatropha oil was purchased at the local market in Moga, Punjab, India. Ethanol and sodium hydroxide was purchased from Merck. All other solvent and reagents were of AR grade and were obtained from Merck.

4.2 Preparation of Sodium Ethoxide

100 ml of ethanol was measured and poured into measuring flask by funnel; 5.0g of sodium hydroxide was carefully added to the measuring flask. Mixture of the solution with proper shaking until sodium hydroxide completely dissolved in the ethanol, forming sodium Ethoxide.

4.3 Transesterification Reaction

Used sorghum and Jatropha oil 100 ml and an appropriate value of ethanol (22-28ml) with alkaline catalyst sodium ethoxide (1% - 10%) placed into a dry reaction flask equipped with reflux condenser and magnetic stirrer. Reaction mixture was mixed for 120 minutes at temperature 60°C that is shown figure 3.

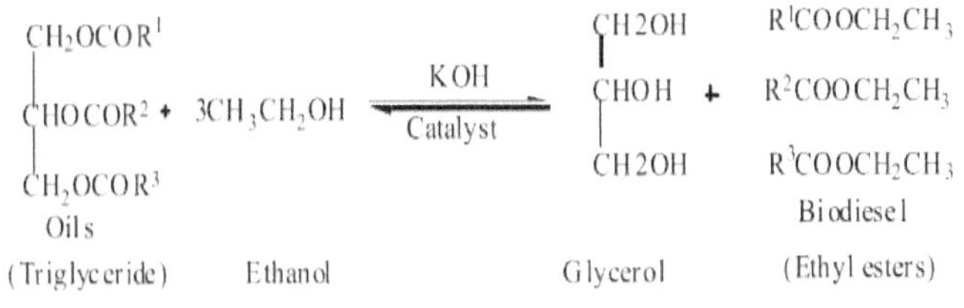

Figure 3: Transesterification Reaction of Sorghum and Jatropha Oil

The ethyl ester (biodiesel) layer was separated from the glycerol layer in a separating funnel. Unreacted oil, ethanol, glycerol, catalyst residue and small amount of produced soap are separated in the separating funnel, this layer was washed with water, until the washing was neutral. This ethyl ester (biodiesel) dried and filtered.

5. RESULT AND DISCUSSION

The experimental results, properties of used sorghum oil are shown in Table 2.

Table-2: Physical and Chemical Properties of sorghum and Jatropha oil.

Properties	Sorghum oil	Jatropha oil
Density (kg/m^3)	0.9099	940
Viscosity (mm^2/s)	34.5	25.5
Specific gravity (g/ml)	0.915	
Acid value (mg NaOH/g)	0.434	
Iodine value (gI^2/100g)	108-122	
Saponification value (mg aOH/g)	181-191	
FFA (mg NaOH/g)	0.864	
Plash point ^{0}C	225	225

Experiments were designed to determine oil quality, temperature, time and amount of alkaline catalyst concentration of reaction affected the yield and properties of biodiesel. The result was analyzed using normal probability method. It indicated that the catalyst concentration was the most important factor affecting the methyl ester (biodiesel).

The catalyst concentration transesterification experiment conducted using the stated reaction parameters the experiment was triplicated and average experimental result evaluated. Results obtained are presented in Table 3.

Table-3: Effect of various catalyst concentration of sorghum biodiesel

Biodiesel Properties	Various catalyst concentration					
	5%	6%	7%	8%	9%	10%
Acid value mgNaOH/g	0.36	0.35	0.34	0.34	0.32	0.32
Iodine value gI^2/100g	98	98	97.5	97.5	97	97
Saponificatio n value mgNaOH/g	170	170	168	168	166	166
Yield of crude biodiesel%	88	88	88.5	89.5	91	90.5
Viscosity (mm^2/s)	3.27	3.27	3.25	3.25	3.24	3.25
Density kg/m^3	0.871	0.871	0.870	0.87	0.87	0.87
Specific gravity g/ml	0.878	0.878	0.876	0.876	0.875	0.875

The methyl ester sorghum oil (MESO) biodiesel prepared using 100 g sorghum oil, different catalyst concentration (1%-10%) at 60 C, 120 minute reaction time, that yield 91% sorghum biodiesel. The maximum yield of sorghum biodiesel is at 9% catalyst concentration. These losses are expected to be some un-reacted alcohol, residual catalyst and emulsion removed during the washing stage of the production. The results stated are averages of three different experimental runs. Detailed results for each of the experimental catalyst concentration are presented Table 3.

5.1 Effect of catalyst concentration on biodiesel production

As a catalyst in the process of alkaline methanol transesterification, sodium methoxide has used in the concentration of 5% (v/v) to10% of oil. The results obtained from methanol transesterification of sorghum oil show that the type of catalyst concentration has an important role.

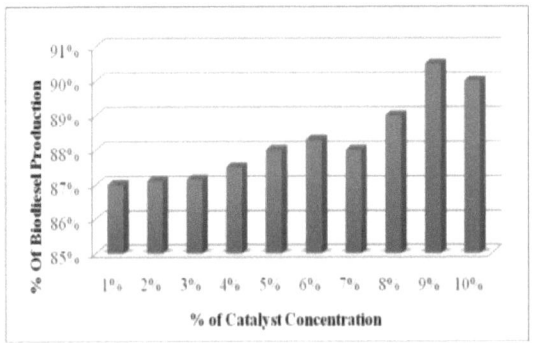

Figure 7: Effect of catalyst concentration on biodiesel production

Namely 9% (v/v) sodium methoxide has given the best yield of biodiesel shown in figure 7. Also eventually soap formation, which appears as a consequence of increased concentration of catalyst, has been avoided.

5.2 Effect of catalyst concentration on viscosity

Viscosity of sample was measured with the help of Red wood Viscometer No.1. Time of gravity flow of fixed value (50 ml) of sample was measured. The experiment was performed at 38/40°C. Below figure 8 shows the increasing catalyst concentration percentage decreasing viscosity but after 9% it remains constant.

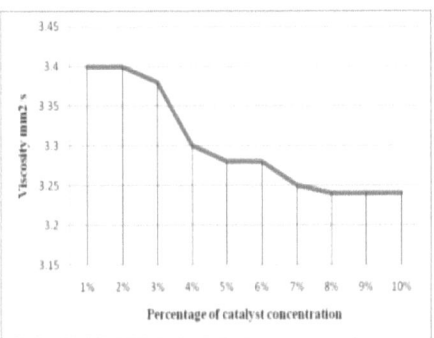

Figure 8: Effect of catalyst concentration on biodiesel viscosity

5.3 *Effect of catalyst concentration on density*

Density was measured using the standard method (BIS, 1972), capillary stopper relative density bottle of 50 ml capacity were used to determine density of biodiesel. Below figure 9 shows the increasing catalyst concentration percentage decreases density but after 9% it remains constant.

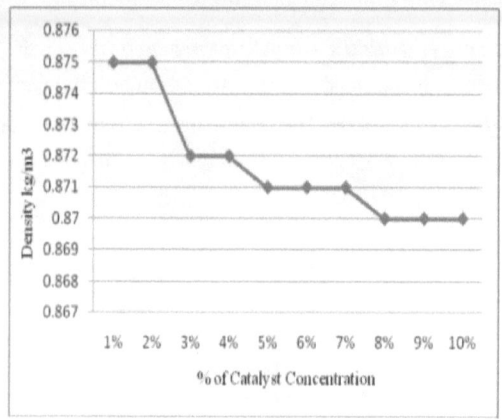

Figure 9: Effect of catalyst concentration on biodiesel viscosity

6. CONCLUSION

The sorghum and jatropha biodiesel obtained in this way can be used as fuel in diesel engines, because of satisfying properties that could be compared with standard biodiesel. To lower price and to make biodiesel competitive with petro-diesel, less expensive feedstock such as sorghum oil have to use in its production. Biodiesel of good quality can be produced from sorghum and jatropha oil in subsequent reaction conditions; 9% (v/v) sodium ethoxide, temperature at 60⁰C and time 120 minutes. The increase of the quantity of catalyst does not contribute to the growth of the yield and the quality of the biodiesel. Ethanol transesterification of sorghum oil and production of biodiesel is another possibility for producing cheap alternative fuels, which could reduce pollution, recycle CO_2 and protect the environment.

7. ACKNOWLEDGMENT

The authors are extremely grateful to the Director, School of Management Sciences Lucknow, and Shivrampati Agro-chemicals Lucknow, Uttar Pradesh, India for providing the valuable facilities.

REFERENCES

[1] G. Best, "Alternative Energy crop for Agriculture Machinery Biofuel- Focus on Biofuel", Agricultural Engineering International: the CIGR Ejournal. Invited Overview 2006, 8, 13.

[2] S. Romano, "Plant oils. A new alternative plant oil fuels. Proc. International Conference on Plant and Plant Oils as Fuels", *American Society of Agricultural Engineers* 1982,4 (82), pp, 101-105

[3] M. Shyam, "Investigations on the characteristics and use of some plant oils as diesel fuel substitute for IC engines. Ph.D", Thesis. Department of Farm Power and Machinery, PAU, Ludhiana, India-141001, 1984.

[4] R.S. Dass and J. Prasad, "Agricultural Engineering Today" 1978, Vol. 2 (2), pp, 14.

[5] A.K. Jain, "Thermo-chemical conversion of biomass", Porc. National conference, PAU, Ludhiana, 30-31 October, 1987.

[6] W. Bandel, "Proc. Of International Symposium on Alcohol Fuel Technology-Methanol and Ethanol" Wolfsburg, Federal Republic of Germany, 1977.

[7] A.W. Schwab, G. J. Dykstra, E. Selke, S. C. Sorenson and E. H. Pryde, "Dieselfuel from thermal decomposition of soybean oil," J Am Oil Chem Soc., 1988, Vol. 65, pp. 1781-1786.

[8] T. Samukawa, M. Kaieda, T. Matsumoto, K. Ban, A. Kondo, Y. Shimada, H. Noda and H. Fukuda, "Pretreatment of immobilized *Candida Antarctica* lipase for biodiesel fuel production from plant oil," J. Biosci. Bioeng. 2000, Vol. 90, pp 180-183.

[9] K. Belafi Bako, F. Kovacs, L. Gubicza and J. Hancsok," Enzymatic biodiesel production from sunflower oil by Candida antarctica lipase ina solvent-free system," Biocatal Biotransform, 2002, Vol. 20, pp 437-439.

[10] V. Dossat, D. Combes and A. Marty, "Lipase-catalyzed transesterification of high oleic sunflower oil," Enzyme Microbiol Technol, 2002, Vol. 30, pp 90-94.

[11] K. Oznur, M. Tuter and H.A. Aksoy, "Lipase catalyzed production of biodiesel," J. Am..Oil Chem. Soc., 1996, Vol. 73, pp. 1191-1195.

[12] L. A. Nelson, T. A. Foglia and W. N. Marmer,"Lipase catalyzed production of biodiesel", J. Am. Oil Chem. Soc., 1996, Vol. 73, pp 1191-1195.

[13] R. D. Abigor, P. O. Uadia, T. A. Foglia, M. J. Hass, K. C. Jones, E. Okpefa, J. U. Obibuzor and M. E. Bafor, "Lipase-catalyzed production of biodiesel fuel from some Nigerian lauric oils," Biochem . Soc. Trans, 2000, Vol. 28, pp. 979-981.

[14] E. Crabbe, C. Nolasco-Hipolito, G. Kobayashi, K. Sonomoto and A. Ishizaki, "Biodiesel production from crude palm oil and evaluation of butanol extraction and fuel properties," Process Biochem., 2001, Vol. 37, pp. 65-71.

[15]   G. Francis, R. Edinger, and K. Becker, " A concept for simultaneous wasteland reclamation, fuel production, and socio-economic development in degraded areas in India: Need, potential and perspectives of Jatropha plantations," Nat. Resource Forum, 2005, Vol. 29 pp. 12-24.

[16]   FAO, Food and Nutrition series N0-27, 1995.

[17]   Hand book of Agriculture. (ICAR), 5th edition Oct. 2006.

[18]   R. Maiti, "Sorghum Science," New Delhi, India; Oxford and IBH Publishing Co. Pvt. Ltd. 1996, 352.

[19]   www.chempro.in/Fatty acid.htm

6

Transforming Land Usage by Agricultural Engineering - A Study of Saudi Arabia

Ashok Sengupta[1*] and Asad Kareem Usmani[2]

[1,2] Department of Management, School of management Sciences, Luckow

*e-mail: ashokgupta@smslucknow.ac.in

ABSTRACT

Saudi Arabia is 5th largest country in Asia. 95% of country is covered by sand and infertile land. It is one of six countries having no permanent river and average rainfall is 4 inches annually. Six decades before the country was depends on other country to fulfil its demand of food, vegetable and dairy items. Since then, government was initiated new economic policy for agricultural development. It provides fund for research and development in the field of agriculture engineering which developed innovative idea to produce food from desert. In 1990, the country was one of the largest producer of Wheat in the world. This paper based on descriptive research on secondary data to study agricultural project in Saudi Arabia in last 60 years. Analysis of data is by using graphical representation.

Keyword:*Economic Policy, Agricultural Engineering, Saudi Arabia.*

1. INTRODUCTION

In the dynamic tapestry of our planet, the transformation of land usage stands as a testament to the evolving interplay between human societies and the environment. As we navigate the 21st century, the pressing need for sustainable land management practices has become increasingly evident, driven by the intricate challenges posed by climate change, population growth, and urbanization. This book chapter embarks on a profound journey into the realm of "Land Usage Transformation," delving into the multifaceted dimensions of environmental changes and the quest for sustainable practices.

Our exploration begins with a reflective examination of the historical trajectory of land usage, tracing the roots of human interaction with the land from ancient civilizations to the present day (Diamond, 2005; McNeill, 2003). Drawing from a diverse array of disciplines, including ecology, geography, sociology, and environmental science, this chapter seeks to unravel the complex web of factors influencing land usage patterns. By examining the historical context, we aim to uncover valuable insights that can inform our understanding of the current state of global landscapes and the challenges that lie ahead.

Amidst the challenges, however, lies a profound opportunity for positive change. The latter sections of this chapter will navigate through contemporary approaches and innovative strategies aimed at fostering sustainable land usage. From regenerative agriculture and urban planning to conservation initiatives and technological advancements, we will explore the promising avenues that hold the potential to harmonize human needs with ecological integrity.

Throughout our exploration, we will weave a narrative enriched with empirical evidence and case studies, ensuring a comprehensive understanding of the diverse landscapes undergoing transformation. Additionally, this chapter will emphasize the importance of interdisciplinary collaboration in addressing the complexities associated with land usage transformation, underscoring the need for a holistic and inclusive approach.

As we embark on this intellectual journey, let us navigate the intricate terrain of land usage transformation with an open mind and a commitment to cultivating a sustainable future. By synthesizing knowledge from various disciplines and harnessing the wisdom embedded in our shared history, we aspire to contribute to the ongoing dialogue surrounding the responsible stewardship of our planet's precious landscapes.

2. AGRICULTURAL ENGINEERING

Agricultural engineering stands at the forefront of technological innovation, playing a pivotal role in transforming the global landscape of food production. This interdisciplinary field intersects engineering principles with agricultural sciences, aiming to address the challenges posed by an ever-growing global population, climate change, and the imperative for sustainable resource management. In this exploration of agricultural engineering, we delve into the diverse facets of this field, examining its evolution, key methodologies, and its crucial contributions to the development of sustainable food systems.

2.1 Evolution and Historical Context

The roots of agricultural engineering trace back to the mechanization of farming practices during the Industrial Revolution. However, it has evolved significantly over the years, incorporating cutting-edge technologies and practices to enhance productivity, efficiency, and environmental sustainability (Sharma, 2018). From the development of advanced machinery to precision farming techniques, agricultural engineering has played a crucial role in modernizing agriculture and meeting the demands of a growing population.

2.2 Key Methodologies and Technological Advancements

Agricultural engineering encompasses a wide array of methodologies and technologies designed to optimize various aspects of farming. Precision agriculture, for example, utilizes information technology and satellite imagery to enhance decision-making processes related to crop management, irrigation, and resource utilization (Srinivasan et al., 2017). The integration of robotics and automation in agriculture is another frontier, revolutionizing tasks such as planting, harvesting, and monitoring crop conditions (Hemming et al., 2020). These advancements not only increase efficiency but also contribute to the sustainable use of resources.

2.3 Contributions to Sustainable Agriculture

The role of agricultural engineering in promoting sustainability cannot be overstated. Sustainable farming practices, such as conservation tillage, water management systems, and organic farming techniques, are integral components of agricultural engineering strategies (Alam et al., 2021). Through innovative irrigation systems and crop modeling, agricultural engineers contribute to water conservation and efficient use, addressing the challenges of water scarcity in many regions (Kisekka et al., 2018). Moreover, the development of post-harvest technologies ensures minimal food wastage, further aligning agricultural engineering with the principles of sustainability (Mittal et al., 2019).

2.4 Challenges and Future Directions

While agricultural engineering has made significant strides, challenges persist. Issues such as adapting to climate change, ensuring food security, and addressing the socio-economic implications of technological advancements are critical considerations for the field (Rosenzweig et al., 2014). The future of agricultural engineering lies in the continued integration of emerging technologies, interdisciplinary collaboration, and a commitment to addressing global challenges.

3. SAUDI ARABIA AND ITS CLIMATE

Saudi Arabia is a vast and predominantly arid country located on the Arabian Peninsula in the Middle East. The country has a diverse climate, characterized by extreme temperatures, low precipitation, and varied geography. Here are key aspects of Saudi Arabia's climate:

3.1 Arid Climate

Saudi Arabia experiences a predominantly arid climate, with large areas characterized by desert conditions. The Rub' al Khali, also known as the Empty Quarter, is one of the largest continuous sand deserts in the world, covering much of the southern part of the country.

3.2 High Temperatures

The temperatures in Saudi Arabia can be extremely high, especially during the summer months. In the central and eastern regions, temperatures often exceed 40°C (104°F), and in some areas, they can reach even higher levels. Coastal areas generally experience milder temperatures due to the influence of the Red Sea and the Arabian Gulf.

3.3 Temperature Extremes

There can be significant temperature variations between day and night. While daytime temperatures can be scorching, nighttime temperatures can drop considerably, providing some relief. In winter, temperatures in the northern and central regions may drop below freezing, especially during the night.

3.4 Limited Precipitation

Saudi Arabia is characterized by low annual rainfall. Most parts of the country receive less than 100 mm (4 inches) of rain annually, making it one of the driest regions on Earth. The western mountains, including the Asir region, receive somewhat higher precipitation due to orographic lifting.

3.5 Seasonal Winds

The country experiences seasonal winds, such as the Shamal, a northwesterly wind that brings dust and sandstorms, particularly during the summer months. The Khamsin, a hot and dry wind, can also occur, impacting visibility and contributing to the arid conditions.

3.6 Red Sea and Arabian Gulf Influence

Coastal areas along the Red Sea and the Arabian Gulf experience a more moderate climate compared to inland regions. The presence of these water bodies mitigates temperature extremes, and coastal cities like Jeddah and Dammam benefit from milder conditions.

3.7 Geographic Variation:

Saudi Arabia's geography varies from vast deserts to mountainous regions. The Asir Mountains in the southwest receive more rainfall and have a more temperate climate, supporting different flora and fauna compared to the arid lowlands.

3.8 Water Scarcity

The arid climate and low precipitation contribute to water scarcity in Saudi Arabia. The country has implemented various water conservation measures, including advanced irrigation techniques and desalination plants to address the demand for water in agriculture and urban areas.

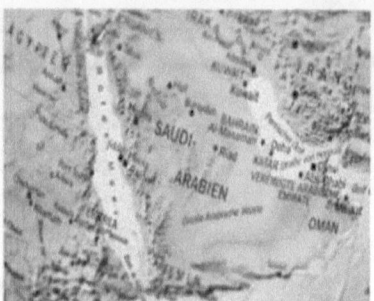

Figure 1 : Source : Common Creative License Geographical Map of Saudi Arabia)

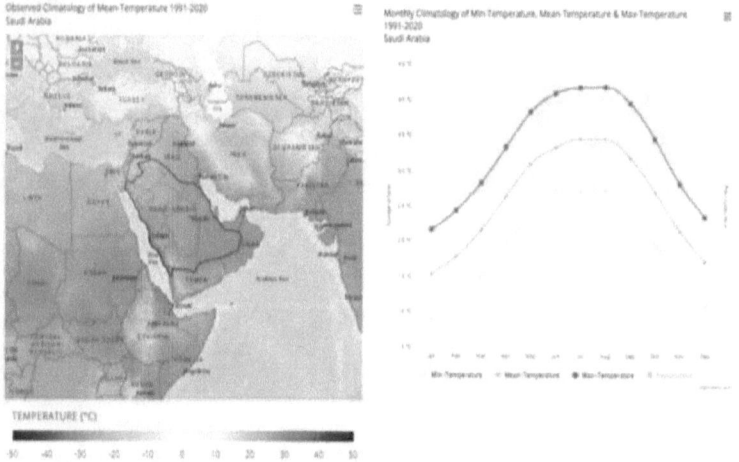

Figure 2 : Climate Map of Saudia

4. CIRCULAR AGRICULTURE AND ITS CONCEPT IN SAUDI ARABIA

Circular agriculture, also known as circular farming or circular economy in agriculture, is an approach that seeks to minimize waste and make the most of available resources by closing loops within the agricultural system. This concept aligns with sustainability principles and focuses on regenerating natural systems, optimizing resource use, and reducing environmental impacts. In the context of Saudi Arabia, where water scarcity and arid conditions pose significant challenges to agriculture, circular agriculture becomes particularly relevant. Here are key aspects of the concept and its application in Saudi Arabia:

4.1 Resource Efficiency

Circular agriculture emphasizes the efficient use of resources, including water, nutrients, and energy. In Saudi Arabia, where water resources are limited, adopting technologies and practices that minimize water consumption in agriculture becomes crucial. Precision irrigation, soil moisture monitoring, and water recycling are examples of resource-efficient approaches.

4.2 Waste Reduction and Recycling

Circular agriculture aims to reduce waste and promote the recycling of by-products. In Saudi Arabia, this could involve the responsible management of agricultural residues, such as crop residues and organic waste. Utilizing these residues for composting or bioenergy production contributes to closing nutrient cycles and reducing environmental impact.

4.3 Closed Nutrient Loops

Circular agriculture focuses on closing nutrient loops by recycling and reusing nutrients within the farming system. This involves practices such as crop rotation, cover cropping, and the use of organic fertilizers. In Saudi Arabia, where soil fertility is a concern, sustainable nutrient management is essential for maintaining agricultural productivity.

4.4 Precision Agriculture Technologies

Implementing precision agriculture technologies is a key component of circular agriculture. In Saudi Arabia, adopting precision farming techniques, such as GPS-guided machinery, drones, and sensor technologies, can enhance the efficiency of resource use, reduce input waste, and optimize crop yields.

4.5 Agroecology and Biodiversity

Circular agriculture encourages agroecological practices that work in harmony with natural ecosystems. Integrating biodiversity into agricultural landscapes, using agroforestry techniques, and supporting natural pollinators are strategies that contribute to ecological balance. This can be particularly important in arid regions like Saudi Arabia.

4.6 Renewable Energy Integration

Circular agriculture promotes the use of renewable energy sources within the farming system. In Saudi Arabia, where sunlight is abundant, integrating solar energy for irrigation, processing, and other agricultural activities aligns with circular economy principles and contributes to sustainability.

4.7 Technology Adoption and Innovation

Embracing technological innovation is essential for the successful implementation of circular agriculture. In Saudi Arabia, investments in research and development, as well as the adoption of smart farming technologies, can drive innovation in sustainable agricultural practices.

4.8 Policy Support and Collaboration

Circular agriculture requires a supportive policy environment and collaboration between stakeholders. Governments, farmers, researchers, and industries in Saudi Arabia can work together to create policies that incentivize circular practices, provide financial support, and promote knowledge exchange.

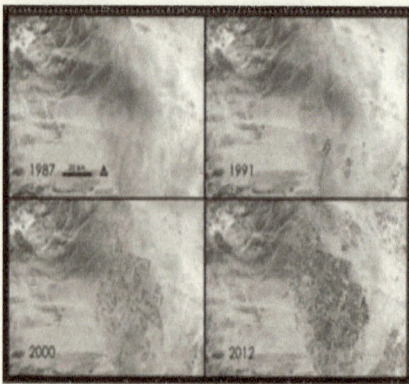 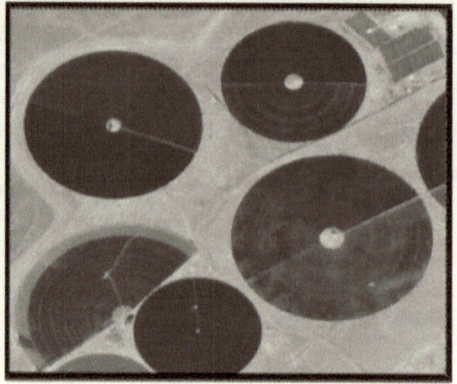

Figure 3: Satellite Image of Change in Landscape of SA

Figure 4: Circular Farming in Saudi Arabia (Satellite Picture)

Source : https://www.flickr.com/photos/gsfc/7027523783/in/set-72157623424324229

5. APPLICATION OF AGRICULTURE ENGINEERING FOR LAND USAGE

5.1 Precision Agriculture Techniques

Agricultural engineers in Saudi Arabia implement precision agriculture techniques to optimize resource use. This includes the use of advanced technologies such as GPS-guided tractors and drones for crop monitoring. Precision agriculture helps in efficient water management, fertilizer application, and pest control, leading to improved crop yields with minimal environmental impact.

Figure 5: AI based Farming

5.2 Drip Irrigation and Water Management

Given the scarcity of water in the region, Saudi Arabia has invested heavily in advanced irrigation systems. Drip irrigation, in particular, is widely used to deliver water directly to the plant roots, minimizing water wastage. Additionally, agricultural engineers have developed innovative water management systems that enhance water use efficiency and reduce the overall demand for this precious resource.

Figure 6: Use of Technology in farming

5.3 Greenhouse Farming

To mitigate the extreme climatic conditions, greenhouse farming has gained popularity in Saudi Arabia. Agricultural engineers design and implement controlled environment systems within greenhouses, allowing for year-round cultivation of crops. This approach helps to shield plants from excessive heat, sandstorms, and water evaporation, creating a more favorable and sustainable growing environment.

Figure 7: : Greenhouse Farming

5.4 Desalination Technology

Addressing the water scarcity challenge, Saudi Arabia has invested significantly in desalination technologies. Agricultural engineers use desalinated water for irrigation in certain areas, allowing for agricultural activities in regions where freshwater resources are limited. This technology helps expand arable land and supports the cultivation of crops that are crucial for food security.

Figure 8: Vertical Farming

5.5 Crop Selection and Biotechnology

Agricultural engineers collaborate with agronomists and biotechnologists to develop crops that are well-suited to the arid conditions of Saudi Arabia. This involves genetic modifications and breeding programs to create drought-resistant and heat-tolerant crop varieties. By selecting and developing crops that thrive in such environments, the country can optimize land usage and enhance agricultural productivity.

5.6 Aquaponics and Hydroponics

In pursuit of sustainable and efficient land usage, Saudi Arabia has embraced aquaponics and hydroponics systems. These soilless cultivation methods, designed by agricultural engineers, allow for the cultivation of crops with minimal water consumption. Aquaponics, which combines aquaculture with hydroponics, is especially useful in maximizing resource utilization while minimizing environmental impact.

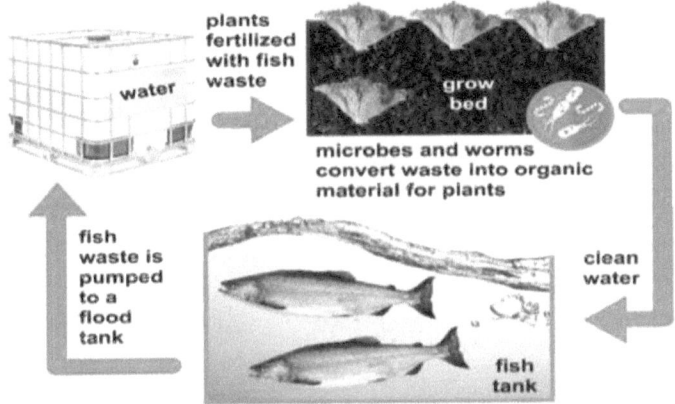

Figure 9: Aquaponics Farming

Figure 10: Hydroponic Farming

5.7 Research and Innovation

Saudi Arabia invests in research and innovation in agricultural engineering to stay at the forefront of sustainable land usage practices. Ongoing research projects focus on developing new technologies, improving existing methodologies, and finding innovative solutions to the unique challenges posed by the country's climatic conditions.

6. CONCLUSION

Understanding and adapting to the unique climatic conditions of Saudi Arabia is crucial for various sectors, including agriculture, water resource management, and infrastructure development. The country continues to invest in innovative solutions to address challenges posed by its climate while striving for sustainability and resilience.

In conclusion, agricultural engineering stands as a dynamic and essential field, driving innovations that shape the future of food production. From historical mechanization to contemporary precision agriculture, this discipline plays a crucial role in advancing sustainable farming practices. As we confront the complexities of feeding a growing population in the face of climate change, agricultural engineering remains a beacon of hope, offering technological solutions that can transform our agricultural systems into models of resilience and efficiency.

Saudi Arabia relies on agricultural engineering to overcome the challenges associated with arid land and water scarcity. Through the implementation of precision agriculture, advanced irrigation systems, greenhouse farming, desalination, biotechnology, and innovative cultivation methods, the country aims to optimize land usage, ensure food security, and promote sustainable agricultural practices.

REFERENCES

[1]   Addas, A. (2020). Enhanced Public Open Spaces Planning in Saudi Arabia to Meet National Transformation Program Goals. Current Urban Studies, 8(02), 184.

[2]   Alam, M. M., Dutta, D., & Mallick, D. L. (2021). Sustainable Agriculture. Boca Raton, FL: CRC Press.

[3]   Center, N. G. S. F. (n.d.). NASA Sees Fields of Green Spring up in Saudi Arabia. Flickr. https://www.flickr.com/photos/gsfc/7027523783/in/set-72157623424324229

[4]   Diamond, J. M. (2005). Collapse: How Societies Choose to Fail or Succeed. New York: Penguin Books.

[5]   Hemming, J., Blackmore, B. S., van den Bosch, C. K., & Wien, H. C. (2020). Agricultural Automation: Fundamentals and Practices. Boca Raton, FL: CRC Press.

[6]   Kisekka, I., Moriasi, D., Gowda, P. H., & Steiner, J. L. (2018). Advances in Irrigation and Drainage: Sustainable Strategies and Systems. Boca Raton, FL: CRC Press.

[7]   McNeill, J. R. (2003). Something New Under the Sun: An Environmental History of the Twentieth-Century World. New York: W. W. Norton & Company.

[8]   Mittal, A., Gupta, A., & Singh, D. (2019). Post-harvest Technologies for Fruits and Vegetables. Boca Raton, FL: CRC Press.

[9]   Rosenzweig, C., Elliott, J., Deryng, D., Ruane, A. C., Müller, C., Arneth, A., ... & Jones, J. W. (2014). Assessing agricultural risks of climate change in the 21st century in a global gridded crop model intercomparison. Proceedings of the National Academy of Sciences, 111(9), 3268-3273.

[10]   Saudi Vision 2030. (n.d.). https://www.vision2030.gov.sa/en/

[11]   Sharma, R. (2018). Agricultural Mechanization and Sustainable Agriculture. Boca Raton, FL: CRC Press.

[12]   Srinivasan, A., McKinion, J. M., Crosson, P., & Ma, L. (2017). Agricultural decision-making using remotely sensed data. Boca Raton, FL: CRC Press.

[13]   The National Research and Development Center for Sustainable Agriculture | Estidamah. (2020, December 22). Estidamah. https://estidamah.gov.sa/en

[14]   World Bank Open Data. (n.d.). World Bank Open Data. https://data.worldbank.org/topic/19

7

Analysis of Physico-Chemical Characteristics of Water Quality Discharge from CETP, Banthar, Unnao

Sanjeev Kumar Pandey[1*], Asha Kulshrestha[2] and Sudhaker Dixit[3]

[1,2] Department of Civil Engineering, School of management Sciences, Luckow

Department of HAS, School of management Sciences, Luckow

*e-mail: sanjeevkp3@gmail.com

ABSTRACT

In this research work my plan has been to study the analysis of physico-chemical characteristics of Water Quality Discharged from CETP, Banthar, Unnao. Total 8 water quality parameters such as- (i) Avg. Flow/Day, (ii) Avg. pH of the day at final, (iii) D.O. (CAT), (iv) D.O. (EAT), (v) COD Final outlet, (vi) BOD Final outlet. (vii) SS(mg/L), & (viii) Chromium (Raw & Final) have been studied related to water quality discharged from CETP, Banthar, Unnao.

This report has been prepared to study the quality of water discharged from Common Effluent Treatment Plant (CETP), Banthar, Unnao. The water received from different tanneries collected at CETP, Banthar, Unnao for treatment. Various unit operations- Barscreen, grit chamber, equalization tank, flash mixer (lime alum), clariflocculator, 1st stage aeration tank (CAT) & 2nd stage aeration tank (EAT) are involved in the treatment of tannery effluent. The main aim of this study is to use of this treated water for agricultural purpose.

Key Words: *Water quality, Water quality parameters, Avg. Flow/Day, pH, DO, COD, BOD, SS, Chromium, Effluent, tanneries & CETP.*

1. INTRODUCTION

Water is a precious resource. Most of the earth water is sea/salty water (about 97.5%) and about 2.5% of the water is fresh water that does not contain significant levels of dissolved minerals or salt and 2/3rd of that is frozen in ice caps and glaciers. Only 0.01% of the total water of the planet is accessible for consumption. Clean drinking water is a basic human need. Unfortunately, more than one in six people still lack reliable access to this precious resource in developing world.

India accounts for 2.45% of land area and 4% of water resources of the world but represents 16% of the world population. With the present population growth-rate (1.9 per cent per year), the population is expected to cross the 1.5 billion mark by 2050. The Planning Commission, Government of India has estimated the water demand increase from 710 BCM (Billion Cubic Meters) in 2010 to almost 1180 BCM in 2050 with domestic and industrial water consumption expected to increase almost 2.5 times. The trend of urbanization in India is exerting stress on civic authorities to provide basic requirement such as safe drinking water, sanitation and infrastructure. The rapid growth of population has exerted the portable water demand, which requires exploration of raw water sources, developing treatment and distribution systems.

The raw water quality available in India varies significantly, resulting in modifications to the conventional water treatment scheme consisting of aeration, chemical coagulation, flocculation, sedimentation, filtration and disinfection. The backwash water and sludge generation from water treatment plants are of environment concern in terms of disposal. Therefore, optimization of chemical dosing and filter runs carries importance to reduce the rejects from the water treatment plants. Also there is a need to study the water treatment plants for their operational status and to explore the best feasible mechanism to ensure proper drinking water production with least possible rejects and its management. With this backdrop, the Central Pollution Control Board (CPCB), studied water treatment plants located across the country, for prevailing raw water quality, water treatment technologies, operational practices, chemical consumption and rejects management.

2. METHODOLOGY

pH is the negative $\log_{10}$ of the hydrogen ion concentration in a solution. It can be measured by colorimetric methods using various indicator or paper strips. However, the uses of colorimetric methods are less convenient and less accurate. For accurate measurement of pH, electrometric methods are used employing the hydrogen ion sensitive electrodes.

2.1 Apparatus Used in The Analysis

There are a number of makes and models available for pH meters. Portable pH meters operated by battery can also be obtained. The accuracy of pH can vary from 0.01 to 0.1 depending on the make. Some pH meters employ two electrodes, while others may have a combined glass and reference electrodes. Most pH meters also have a temperature compensation system to avoid the differences arising due to the different temperature.

2.2 Procedure Followed in The Analysis

Follow the instructions given by the manufactures to use the pH meter. Essential aspect to use all the pH meters is to calibrate it with suitable buffers. Ready buffers of pH values are also available in the market. See the pH meter with a buffer whose value is near to the expected pH of the samples. Buffers of different pH values can be made in the laboratory.

A. Potassium hydrogenphthalate buffer (pH = 4)

B. Phosphate buffer (pH = 7)

C. Borax buffer (pH = 9)

2.2.1 Procedure Followed in The Analysis

1. Fill the sample in a glass stoppered bottle (BOD Bottles of known) volume (100 – 300ml), carefully, avoiding any kind of bubbling and trapping of the air bubbles in the bottle after placing the stopper.

2. Pour 1ml. of each $MnSO_4$ and alkaline KI solutions (in case, the volume of the sample is about 300ml, instead of 1ml of reagents add 2ml solutions of each), well below the surface from the walls. The reagents can also be poured at the bottom of the bottle with the help of special pipette syringes to ensure better mixing of the reagents with the sample. Use always, separate pipettes for these two reagents. A precipitate will appear.

3. Place the stopper and shake the contents well by inverting the bottle repeatedly. Keep the bottle for some time to settle down the precipitate. If the titration is to be prolonged for few days, keep the sample at this stage with the precipitate.

4. Add 1 – 2 ml of concentrated H_2SO_4 and shake well to dissolve the precipitate.

5. Remove either the whole contents, or a part of them (50-100 ml) in conical flask for titration. Prevent any bubbling to avoid further mixing of oxygen.

6. Titrate the contents, within one hour of dissolution of the precipitate against sodium thiosulphate solution using starch as an indicator. At, the end point, initial dark blue colour changes to colourless.

2.2.2 Calculation

When whole contents have been titrated:

$$\text{Diss. Oxygen, mg}/l = \frac{\text{(ml. x N)of titrant x 8 x 1000}}{V_1 - v}$$

When only a part of the contents has been titrated:

$$\text{Diss. Oxygen, mg}/l = \frac{\text{(ml. x N)of titrant x 8 x 1000}}{V_2\left(\dfrac{V_1 - v}{V_1}\right)}$$

Where, V_1 = volume of sample bottle after placing the stopper.

V_2 = volume of the part of the contents titrated.

v = volume of $MnSO_4$ and KI added.

2.2.3 Azide Winkler's Iodometric Method : The presence of certain oxidizing and reducing materials may effectively interference with the determination of oxygen by converting iodide ions to iodine or vice-versa. The azide modification removes the interference of such substances especially nitrite. Nitrite is destroyed by sodium azide (NaN_3). The method, therefore, is suitable particularly in polluted waters, biologically treated waters and in BOD samples.

2.2.4 Reagents used in the analysis : Sodium thiosulphate, 0.025N; Alkaline iodide azide solution, Manganous sulphate solution, starch solution & sulphuric acid.

2.2.5 Procedure followed in the analysis : Follow the same Winkler's method except that; add alkaline iodide azide solution instead of alkaline potassium iodide in the same quantities.

2.2.6 Calculation : Same as for Winkler's iodometric method.

2.3 Measurement of Chemical Oxygen Demand (COD), (APHA, 1998).

Chemical Oxygen Demand (COD) is the measure of Oxygen consumed during the oxidation of the oxidizable organic matter by a strong oxidizing agent. Potassium Dichromate in the presence of sulphuric acid is generally used as an oxidizing agent in determination of COD.

The sample is refluxed with $K_2Cr_2O_7$ and H_2SO_4 in presence of mercuric sulphate to neutralize the effect of chlorides, and silver sulphate (catalyst). The excess of potassium dichromate is titrated against ferrous ammonium sulphate using ferroin as an indicator. The amount of $K_2Cr_2O_7$ used is proportional to the oxidizable organic matter present in the sample.

2.3.1 Reagents Used in The Analysis : Potassium dichromate solution, 0.25 N, Potassium dischromate solution, 0.025 N, Ferrous ammonium sulphate, 0.1 N, Ferrous ammonium sulphate, 0.01 N, Ferroin indicator, Sulphuric acid (H_2SO_4, conc), Mercuric sulphate (HgSO4, solid) & Silver sulphate (Ag_2SO_4, solid)

2.3.2 Procedure followed in the analysis

1. Take 20ml of sample in a 250 – 500ml COD flask. (round bottom or Erlenmeyer flask with a ground joint for Liebig reflux condenser.

2. If the sample is expected to have COD more than 50 mg/l, add 10ml of 0.025 N potassium dichromate solution. Extreme care should be taken in case of low COD samples. A small trace of organic matter in glassware may contribute a significant error.

3. Add a pinch of Ag_2SO_4 and HgSO4. If the sample contains chlorides in higher amount, HgSO4 is added in the ratio of 10: 1, to the chlorides. COD cannot be determined accurately if the sample contains more than 2000 mg/l of chlorides.

4. Add 30ml of sulphuric acid.

5. Reflux at least for 2 hours on a water bath or a hot plate. Remove the flasks, cool and add distilled water to make the final volume to about 140 ml.

6. Add 2-3 drops of ferroin indicator, mix thoroughly and titrate with 0.1 N ferrous ammonium sulphate (with 0.01 N ferrous ammonium sulphate if 0.025 N $K_2Cr_2O_7$ has been used.

7. Run a blank with distilled water using same quantity of the chemicals.

2.3.3 Calculation

COD, mg/l = (b-a) x N of $K_2Cr_2O_7$ x 1000 x 8/ml sample

Where   a= ml of titrant with sample

b= ml of titrant with blank

3. RESULT AND DISCUSSION

Results of the present study are discussed under the following headings-

3.1 Data & Data Collection

The capacity of CETP, Banthar, Unnao is 4.5 MLD (4500 KLD) and an avg. inflow per day (Avg. flow/ D) is about 2.5-3.5 MLD (2500-3500KLD). The avg. value of Water sample (i.e. tannery effluent) collected at CETP, Banthar, Unnao is shown intable 3.1.

Table-3.1: Avg. value of Water sample (i.e. tannery effluent) collected at CETP, Banthar, Unnao

Water Quality Parameter	Avg. Value
Avg. Flow/D (Inflow)	2500-3500 KLD (2.5-3.5 MLD)
pH	7.9-8.5
D.O.(CAT)	2.1-2.2
D.O.(EAT)	1.9-2.0
C.O.D.	2500-3500
B.O.D.	1100-1300
SS	600-700
Chromium (Raw)	1.63-2.82

In Table: All parameters having unit mg/l (except pH & Avg. flow/D)

All datas observed in different treatment processes in different months (June, July & August) in 2011 have been shown in Appendix-A, B& C and have been plotted in different figures (i.e. from figure 3.1 to figure 3.3)

3.2 Data Analysis

Following variables are given in Table 3.2, 3.3 & 3.4. We have used Maple software to present the data in every month in a graphical form in different figures. (i.e. from Figure 3.1 to Figure 3.3).

Variations in values of different water quality parameters have been observed because in various treatment processes either there is more or less supply of inflow or more or less supply of various chemicals as a result value of water quality parameter is more or less.

3.2.1 Avg. Flow/D : It is actually the inlet (inflow) per day expressed in KLD. In Figure 3.1, the Avg.Flow/D has been plotted for the month of June-2011. In the case of CETP, Banthar, Unnao, from the graph it is noted that the Avg. Flow is lowest (min.) on 13.06.11 (i.e.2400 KLD) and the other lowest value on 14.06.11 (i.e. 2450 KLD). It appears that on these dates different tanneries attached to CETP, Banthar, Unnao were not functional and because of which the flow of water is lower.

It is further noted that apart from these two extreme cases the Avg. flow is 3117 KLD & max. value 3300 KLD is noted on 21.06.11,24.06.11,25.06.11,28.06.11 & 30.06.11. (See Figure 3.1 & Table 3.2).

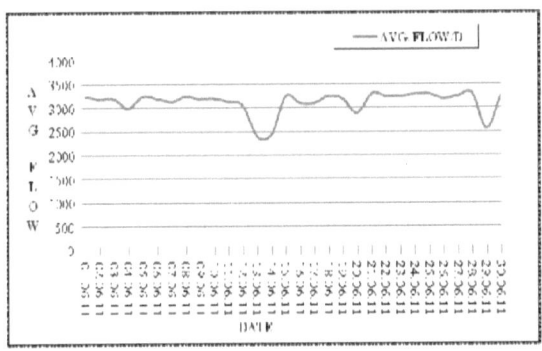

Figure 3.1 : Avg. flow/D of June-2011 at CETP, Banthar, Unnao

In the month of July, Figure 3.2 shows the Avg. flow is 3248 KLD; while lowest (min.) value 2050 KLD is noted on 20.07.11; and max.value 3650 KLD is noted on 29.07.11.
(See Figure 3.2 & Table 3.3).

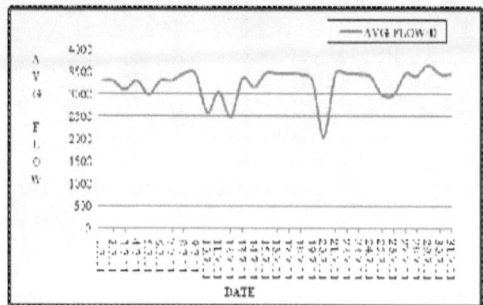

Figure 3.2 : Avg. flow/D of July-2011 at CETP, Banthar, Unnao

In the month of August, Figure 4.3 shows the Avg. flow is 3144 KLD; while lowest (min.) value 1550 KLD is noted on 16.08.11 and max. value 3700 KLD is noted on 09.08.11,10.08.11,11.08.11 & 12.08.11.(See Figure 3.3 & Table 3.4).

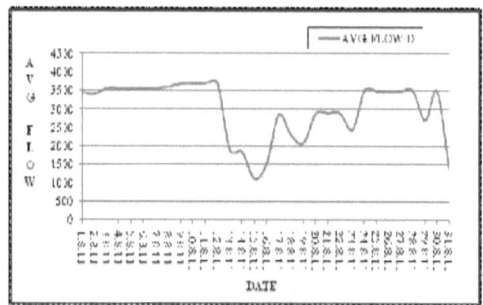

Figure 3.3: Avg. flow/D of Aug-2011 at CETP, Banthar, Unnao

In Figure 3.1, 3.2 & 3.3; it appears that on different dates in the month of June, July & August respectively the different tanneries attached to CETP, Banthar, Unnao were not functional and because of which the flow of water is lower. Due to which Avg. flow/D is supplied in low amount as inlet (i.e. from effluent of different tanneries) at CETP, Banthar, Unnao.

Variations in values of avg. flow/day of water (inflow) have been observed because there is more or less supply of effluents from different tanneries on different dates at CETP, Banthar, Unnao.

Table 3.2, 3.3, 3.4 has been shown below-

Table- 3.2 : Parameters testing analysis of CETP Banthar for the month of June-2011

Date	AVG. FLOW/ D (KLD)	AVG. pH of THE DAY AT FINAL	D.O. CAT (mg/l)	D.O. EAT (mg/l)	COD FINAL OUTLET (mg/l)	BOD FINAL OUTLET (mg/l)	SS (mg/l)	CHROMIUM RAW/FINAL (mg/l)
01.06.11	3250	7.5	3.18	2.07	184	......	68	2.0813/0.0761
02.06.11	3200	7.4	2.93	2.18	192	......	71	2.1374/0.0653
03.06.11	3200	7.3	2.73	2.35	200	25	67	2.0873/0.0718
04.06.11	3000	7.4	2.48	2.17	208	......	61	2.07152/0.0618
06.06.11	3200	7.5	2.53	2.38	198	21	65	2.1882/0.0782
07.06.11	3150	7.2	2.46	2.16	220	......	58	2.3719/0.0618
08.06.11	3250	7.4	2.37	2.08	203	......	65	2.1876/0.0915
09.06.11	3200	7.5	2.48	2.13	214	23	71	1.9876/0.0903
10.06.11	3200	7.6	2.73	2.36	220	......	75	2.6317/0.0651
11.06.11	3150	7.4	2.84	2.71	227	......	64	2.8193/0.0651
13.06.11	2400	7.2	2.37	1.86	236		73	2.3714/0.0982
14.06.11	2450	7.6	2.43	2.16	246	24	65	2.0381/0.0872
15.06.11	3250	7.5	2.38	2.53	232	......	71	2.1835/0.0715
16.06.11	3100	7.6	2.63	2.43	224	......	65	2.3918/0.0651
17.06.11	3100	7.5	2.71	2.38	216	21	78	2.4139/0.0937
18.06.11	3250	7.6	2.93	2.42	224	......	75	2.1865/0.0991
20.06.11	2900	7.3	2.86	2.71	216	25	67	2.0817/0.0814
21.06.11	3300	7.5	3.05	2.62	200	......	72	2.3817/0.0842
22.06.11	3250	7.2	3.12	2.48	195	......	75	2.0187//0.0673
23.06.11	3250	7.5	3.76	3.05	208	21	80	1.9817/0.0871
24.06.11	3300	7.7	3.42	2.96	200	......	78	1.8714/0.0915
25.06.11	3300	7.2	3.28	3.02	184	......	73	2.1718/0.0735
27.06.11	3250	7.4	3.63	2.98	168	......	76	2.0143/0.0917
28.06.11	3300	7.2	3.41	2.73	160	23	71	1.9814/0.0671
29.06.11	2550	7.5	3.57	2.68	152	......	79	1.8714/0.0708
30.06.11	3300	7.6	3.08	2.52	184	......	81	2.1719/0.1316
MIN.	2400	7.2	2.37	1.86	152	21	58	1.8714/0.0618
MAX.	3300	7.7	3.76	3.05	246	25	81	2.8193/0.1316
AVG.	3117.30	7.43	2.90	2.47	204.27	22.86	70.92	2.1806/0.0803

Note: No data were collected on Sundays & holidays.

Table- 3.3 : Parameters testing analysis of CETP Banthar for the month of July-2011

Date	AVG. FLOW/ D (KLD)	AVG. pH of THE DAY AT FINAL	D.O. CAT (mg/l)	D.O. EAT (mg/l)	COD FINAL OUTLET (mg/l)	BOD FINAL OUTLET (mg/l)	SS (mg/l)	CHROMIUM RAW/FINAL (mg/l)
01.07.11	3300	7.2	2.98	3.28	204	17	79	1.9372/0.1062
02.07.11	3300	7.3	2.92	3.13	184	.....	65	1.6819/0.1132
04.07.11	3300	7.2	2.68	3.42	168	21	72	1.7609/0.09813
05.07.11	3000	7.6	2.48	2.87	200	.....	58	1.9862/0.0831
06.07.11	3300	7.5	2.51	2.91	184	.....	59	2.0614/0.0715
07.07.11	3300	7.3	2.86	3.18	168	20	52	1.9825/0.0618
08.07.11	3450	7.3	2.76	3.28	152	.....	68	1.6714/0.1083
09.07.11	3450	7.5	2.92	2.73	144	.....	53	1.7420/0.1372
11.07.11	3050	7.3	2.84	2.73	176	.....	64	1.4968/0.1073
12.07.11	2500	7.5	3.16	2.68	208	24	71	1.8105/0.0981
13.07.11	3350	7.6	2.98	2.74	200	.....	68	2.0681/0.0872
14.07.11	3150	7.4	3.68	3.24	200	.....	59	1.9813/0.0917
15.07.11	3450	7.2	3.72	3.41	168	21	65	1.4932/0.0816
16.07.11	3450	7.4	3.51	3.48	152	.....	61	1.7814/0.0917
18.07.11	3450	7.3	3.41	3.26	136	18	58	1.9376/0.1074
19.07.11	3300	7.5	3.62	3.28	160	.....	65	1.6825/0.1072
20.07.11	2050	7.4	3.18	3.1	152	.....	53	1.7253/0.0915
21.07.11	3450	7.6	2.98	2.84	187	19	62	1.8465/0.0715
22.07.11	3450	7.4	2.48	2.67	180	.....	58	1.7918/0.0837
23.07.11	3450	7.3	2.38	2.17	174	.....	65	1.8617/0.0792
25.07.11	3000	7.5	2.98	1.73	187	.....	54	1.6972/0.0815
26.07.11	3000	7.4	2.84	2.93	168	23	62	1.8376/0.0913
27.07.11	3450	7.4	3.14	2.87	219	.....	65	1.6714/0.0817
28.07.11	3400	7.2	3.28	3.05	184	.....	54	1.6172/0.0652
29.07.11	3650	7.3	2.98	2.42	224	21	62	1.9817/0.0895
30.07.11	3450	7.5	2.47	1.98	220	.....	73	1.7165/0.0715
MIN.	2050	7.2	2.38	1.73	136	17	52	1.4932/0.0618
MAX.	3650	7.6	3.72	3.48	224	24	79	2.0681/0.1372
AVG.	3248.07	7.39	2.99	2.90	180.73	20.44	62.5	1.8008/0.0907

Note: No data were collected on Sundays & holidays.

Table- 3.4 : Parameters testing analysis of CETP Banthar for the month of August-2011

Date	AVG. FLOW/ D (KLD)	AVG. pH of THE DAY AT FINAL	D.O. CAT (mg/l)	D.O. EAT (mg/l)	COD FINAL OUTLET (mg/l)	BOD FINAL OUTLET (mg/l)	SS (mg/l)	CHROMIUM RAW/FINAL (mg/l)
01.08.11	3450	7.2	3.16	2.38	168	23	70	1.8437/0.0742
02.08.11	3400	7.6	3.42	2.83	200	......	65	1.7653/0.1072
03.08.11	3550	7.3	3.62	2.92	192	......	61	1.6514/0.0917
04.08.11	3550	7.6	3.42	1.92	200	19	52	1.7652/0.0872
05.08.11	3550	7.4	3.65	2.35	168	......	59	2.3173/0.1137
06.08.11	3550	7.5	3.72	2.41	160	......	54	2.1653/0.1072
08.08.11	3600	7.3	3.65	2.56	168	......	48	2.0813/0.0912
09.08.11	3700	7.4	3.18	2.43	152	23	59	2.0873/0.0836
10.08.11	3700	7.2	3.08	2.53	144	......	68	2.1765/0.0913
11.08.11	3700	7.5	3.26	2.98	136	......	68	2.0735/0.0681
12.08.11	3700	7.4	3.18	2.52	152	22	61	2.1718/0.0695
13.08.11	1900	7.3	3.54	2.67	176	......	58	1.9873/0.0679
16.08.11	1550	7.4	3.51	2.36	136	......	56	1.9817/0.0619
17.08.11	2850	7.6	3.72	2.48	168	20	72	1.9615/0.0693
18.08.11	2300	7.3	3.42	2.37	184	......	62	1.6513/0.0913
19.08.11	2100	7.4	3.47	2.29	188	......	65	1.5419/0.0917
20.08.11	2900	7.2	2.84	2.37	198	24	74	1.4739/0.0871
22.08.11	2900	7.3	2.65	2.37	192	......	62	1.8916/0.0713
23.08.11	2450	7.2	2.41	1.84	216	21	79	1.8172/0.1013
24.08.11	3500	7.4	2.38	1.92	227	......	80	1.9814/0.0615
25.08.11	3500	7.3	2.41	1.86	216	......	64	1.8432/0.0814
26.08.11	3500	7.4	2.62	1.73	208	18	58	2.0812/0.0915
27.08.11	3500	7.3	2.58	1.65	192	......	71	2.1916/0.0816
29.08.11	2700	7.5	2.61	1.82	184	21	68	2.3171/0.1072
30.08.11	3500	7.6	2.48	1.73	168	......	79	1.7159/0.0793
MIN.	1550	7.2	2.38	1.65	136	18	48	1.4739/0.0615
MAX.	3700	7.6	3.72	2.98	227	24	80	2.3173/0.1137
AVG.	3144	7.38	3.12	2.29	179.72	21.22	64.52	1.9414/0.08425

Note: No data were collected on Sundays & holidays.
Similarly all parameters are studies and arranged to evaluate other parameters.

4. CONCLUSION

This report has been prepared to study the quality of water discharged from CETP, Banthar, Unnao. The water received from different tanneries collected at CETP, Banthar, Unnao for treatment. The various unit operated at the treatment plant are as follows- (1) Barscreen, (2) grit chamber, (3) equalization tank, (4) flash mixer (lime alum), (5) clariflocculator, (6) 1 st stage aeration tank (CAT) and (7) 2 nd stage aeration tank (EAT).

The data has been collected of raw & treated water from CETP. The characteristics of the following variables have been studied-

(i) Avg. Flow/D,

(ii) Avg. pH of the day at final,

(iii) D.O. (CAT),

(iv) D.O. (EAT),

(v) COD Final outlet,

(vi) BOD Final outlet,

(vii) SS(mg/L), &

(viii) Chromium (Raw & Final).

REFERENCE

[1]   Annual Report– 2005-06, Central Pollution Control Board, Ministry of Environment & Forests, Govt. of India

[2]   Annual Report–2006-07, Central Pollution Control Board, Ministry of Environment & Forests, Govt. of India

[3]   APHA, Standard Methods for Examination of Water and Wastewater, American Public Health Association, AWWA, WPCF, Washington D. C., (1967)

[4]   APHA, Standard Methods for Examination of Water and Wastewater, American Public Health Association, American Water Works Association & Water Environment Federation, Washington D. C., (1985)

[5]   APHA-AWWA-WPCF, Standard Methods for Examination of Water and Wastewater, American Public Health Association, American Water Works Association & Water Pollution Control Federation, Washington D. C.,(1981)

[6]   Besselievre, E. B.: The Treatment of Industrial Wastes, McGraw-Hill, New York, (1969)

[7]   Clesceri, L.S., Greenberg, A. E. & Eaton, A. D. (eds): Standard Methods for Examination of Water and Wastewater, 20 th ed., APHA; American Public Health Association, Washington D. C., (1998)

[8]   Coppen John, "Advanced Wastewater Treatment Systems", University of Southern Queensland, Faculty of Engineering and Surveying, October (2004)

[9]   De, A. K.: Environmental Chemistry, 6 th ed., New Age International Publishers, New Delhi, (2006), 232-233

[10] Durgananda Singh Chaudhary, Saravanamuthu Vigneswaran, Huu-Hao Ngo, Wang Geun Shim and Hee Moon, "Biofilter in Water and Wastewater Treatment", Korean J. Chem. Eng., 20(6), 1054-1065 (2003)

[11] E. Leghouchi, E. Laib & M. Guerbet, "Evaluation of chromium contamination in water, sediment and vegetation caused by the tannery of Jijel (Algeria): A case study", Environ. Monit. Assess. (2009) 153:111–117

[12] Eckenfelder, W.W., Jr.: Industrial Water Pollution Control, McGraw-Hill, New York, (1966)

[13] "Analysis of Physico-chemical Characteristics of Water Quality Discharged From CETP, Banthar, Unnao" 2011

[14] J. Bunch, V. Madha Suresh and T. Vasantha Kumaran, eds., Proceedings of the Third International Conference on Environment and Health, Chennai, India, 15-17 December (2003). Chennai: Pages 284 – 289

[15] Namasivayam, C. and R. T. Yamuna, "Studies On Chromium (iii) Removal From Aqueous Solution By Adsorption Onto Biogas Residual Slurry and its Application to Tannery Wastewater Treatment", Environmental Chemistry Division, Department Of Environmental Sciences, Bharathiar University Nemero, N. L.: Liquid Wastes of Industry: Theories, Practices, and Treatment, Addison-Wesley, Reading, Mass., (1971)

[16] Peavy, H. S.; Rowe, D. R. & Tchobanoglous, G. Environmental Engineering,McGraw-Hill Book Company, Singapore, (1985), 208-217.

[17] R.K. Sahu, S. Katiyar, Jaya Tiwari and G. C. Kisku, "Assessment of drain water receiving effluent from tanneries and its impact on soil and plants with particular emphasis on bioaccumulation of heavy metals", Journal of Environmental Biology July 2007, 28(3) 685-690 (2007).

[18] Sawyer, C. N., McCarty, P. L. & Parkin, G. F.: Chemistry for Environmental Engineering and Science, 5 th ed., McGraw-Hill, New York, (2003), 593-629.

[19] Sengupta, B., Verma, N.K. , Ansari, P. M. , Paritosh Kumar, Basu D. D., Thirumurthy G. , "Status Of Water treatment Plants In India", Central Pollution Control Board, Ministry of Environment & Forests, (2005)

8

A Review on Solar Hybrid (PV+CSP) Trading System of India to Other Countries

Garima Singh[1*] and Bharat Raj Singh[2]
[1*]BBD National Institute of Technology & Management, Lucknow
e-mail: garima.geeta@gmail.com
[2]School of Management Sciences, Lucknow.

ABSTRACT

Solar power is the key to a clean energy future. Solar power is the conversion of energy from sunlight into electricity, either directly using photovoltaic (PV), indirectly using concentrated solar power, or a hybridization of PV and CSP. This research paper summarizes the geographical area, average DNI, power consumption and solar power production of countries like Morocco, South Africa, Italy, Saudi Arabia etc. including India.

The aim of this research paper is to show that India' solar energy market is growing day by day in comparison to other countries and in near future India's solar suppliers tap down this solar energy and provide this solar energy to international grid for their usage. Carbon free is also possible with solar energy.

Keywords: Photovoltaic, Solar Hybrid, PV and CSP, Concentrated Solar Thermal.

1. INTRODUCTION

Solar energy is used worldwide and is increasingly popular for generating electricity or heating and desalinating water. Solar power plants use one of two technologies- (i) Photovoltaic (PV) systems use solar panels, either on rooftops or in ground-mounted solar farms, converting sunlight directly into electric power and (ii) Concentrated solar power (CSP, also known as "concentrated solar thermal") plants use solar thermal energy to make steam, which is thereafter converted into electricity by a turbine. Many industrialized nations have installed significant solar power capacity into their grids to supplement or provide an alternative to conventional energy sources while an increasing number of less developed nations have turned to solar to reduce dependence on expensive imported fuels. Long distance transmission allows remote renewable energy resources to displace fossil fuel consumption. The geographical area, average DNI, power consumption and solar power production of several countries are summarized as follows:

1.1 Morocco

Geographical area of Morocco is 701,850 km². Morocco has a higher potential of direct and global solar radiation of about 2500kWh/m²/year and 2200kWh/m²/year, respectively, in the southern region [1, 2]. For this reason, Ouarzazate (30°552 083 N, 6°532 363 E) which is located in the south of Morocco receives a high solar radiation. Ouarzazate is also chosen for study bt many researchers. Morocco generates 28,750,440 MWh of electricity as of 2016 (covering 102% of its annual consumption needs),24,229,440MWh produced by non-renewable (fossil fuels: oil, natural gas ,coal) and 4,643,000 MWh by renewable such as hydroelectricity, non-hydroelectric, wind and solar. Morocco consumed 28,250,440 MWh of electricity in 2016 [3].

1.1.1 Solar Power Production in Morocco: Solar power in Morocco is enabled by the country having one of the highest rates of solar insolation among other countries about 3,000 hours per year of sunshine but up to 3,600 hours in the desert. Morocco has launched one of the world's largest solar energy projects costing an estimated $9 billion. The aim of the project is to create 2,000 megawatts of solar generation capacity by the year 2020 [4]. Five solar power stations are to be constructed, including both photovoltaic and concentrated solar power technology. The Moroccan Agency for Solar Energy (MASEN), a public-private venture, has been established to lead the project. The first plant will be commissioned in 2015 [5], and the entire project in 2020 [6]. Once completed, the solar project will provide 38% of Morocco's annual electricity generation.

Loubna Bousselamti et. al [7] discussed the modelling and assessing the performance of hybrid PV-CSP plants in Morocco: A Parametric Study. They concluded that PV-CSP hybridization provides a dispatching energy with a lower LCOE than the CSP alone. PV-CSP hybrid plant not only provide profit in Morocco's economy but it will also very helpful for making ecofriendly community. Ayat-allah Bouramdane et. al [8] discussed the adequacy of Renewable Energy Mixes with Concentrated Solar Power and Photovoltaic in Morocco: Impact of Thermal Storage and Cost. The result show that CSP-TES is the best viable option when it is compared to PV and wind due to reducing the adequacy risk by mitigating variability of renewable energy production. L. Bousselamti et. al [9] discussed the study of hybrid PV-CSP plants considering two dispatch strategies in Ouarzazate. The two dispatching strategies (DS) were used. In dispatch strategy 1 (DS1) PV satisfy the load while the CSP production is used to cover the missing load in daytime and evening and in dispatch strategy 2 (DS2) PV operates in daytime and CSP operates in the evening. The result shows that DS1 is the best choice to minimize the cost of hybrid PV-CSP in Ouarzazate and maximize the annual energy production. Sebastian-James Bode et. al [10] studied retrofitting Operating CSP Plants with PV to Power Auxiliary Loads – Technical Consideration and Case Study. For improving the financial return of CSP projects PV augumentation plays an important role. The paper also discusses the economic potential, benefits and challenges of retrofitting CSP plants with PV to service the parasitic loads of the CSP plant.

1.2 South Africa

The total land area of South Africa is 1,220,813 km² According to the South African Department of Energy, the whole of Africa has sunshine all year round. "The annual 24-hour global solar radiation average is about 220 W/m² for South Africa, compared with about 150 W/m² for parts of the USA, and about 100 W/m² for Europe and the United Kingdom. South Africa has a large energy sector, being the second-largest economy in Africa. The country consumed 227 TWh of electricity in 2018. The vast majority of South Africa's electricity was produced from coal, with the fuel responsible for 88% of production in 2017. As of July 2018, South Africa had a coal power generation capacity of 39 gigawatts (GW). South Africa is planning to shift away from coal in the electricity sector. The country aims to decommission 34 GW of coal-fired power capacity by 2050. It also aims to build at least 20 GW of renewable power generation capacity by 2030. South Africa aims to generate 77,834 megawatts (MW) of electricity by 2030, with new capacity coming significantly from renewable sources to meet emission reduction targets [11].

1.2.1 Solar Power Production in South Africa: Solar power in South Africa includes photovoltaic (PV) as well as concentrated solar power (CSP). In 2016, South Africa had 1,329 MW of installed solar power capacity [12]. Installed capacity is expected to reach 8,400 MW by 2030 [13].

Ahmed Bilal Awan et. al [14] discussed the design and comparative analysis of photovoltaic and parabolic trough based CSP plants. Three different sites of Saudi Arabia (Tabuk, Majmaah & Najran) has been taken for study. Tabuk site is proven to be the best location for both CSP and PV plant because of better average DNI of 7.43 kWh/m²/day. The LCOE of best case CSP plant is 2.73 times higher than that of PV plant and net capital cost of CSP plant is 4.5 times higher compared to PV plant. The capacity utilization factor of the best case CSP plant is 45.4% compared to 30.2 % PV and electrical energy generation of best case CSP plant is 33.3% more compared to best case PV plant.

1.3 Saudi Arabia

Saudi Arabia is a country in Western Asia with a land area of approximately 2,150,000 km² (830,000 sq mi). The direct normal irradiance (DNI) in various regions of the country ranges from approximately 9000 W h/m²/day in the summer months to 5000 W h/m²/day in the winter months. Global horizontal irradiance (GHI) in various regions can be as high as 8.3 kW h/m²/day [15]. Power consumption in Saudi Arabia increased sharply during the 1990–2010 period due to rapid economic development. Peak loads reached nearly 24 GW in 2001 that is 25 times to 1975 level and are expected to approach 60 GW by 2023[16]. The investment needed to meet this demand may exceed $90 billion. Electricity generation is 40% from Oil 52% from Natural Gas and 8% from steam. Generation capacity is approximately 55 GW [17]. A looming energy shortage requires Saudi Arabia to increase its capacity. Capacity is planned to be increased to 120 GW by 2032[17].

1.3.1 Solar Power Production in Saudi Arabia: Solar power in Saudi Arabia has become more important to the country as oil prices have risen. in 2011, over 50% of electricity was produced by burning oil[18]. The Saudi agency in charge of developing the nations renewable energy sector, Ka-care, announced in May 2012 that the nation would install 41 gigawatts (GW) of solar capacity by 2032[19]. It is projected to be composed of 25 GW of solar thermal, and 16 GW of photovoltaics. At the time of this announcement, Saudi Arabia had only 0.003 gigawatts of installed solar energy capacity[20].

Mario Petrollese et. al [21] discussed the optimal design of a hybrid CSP-PV plant for achieving the full dispatchability of solar energy power plants. The results show that PV-CSP hybridization technology is highly cost effective.

1.4. Italy

Italy total area is 301,340 km² (116,350 sq mi), of which 294,140 km² (113,570 sq mi) is land and 7,200 km² is water (2,780 sq mi). In Italy mean annual solar radiation ranges from 3.6 kW per square meter in the Po river plain area, to 4.7 kW per square meter in Central-Southern Italy, to 5.4 kW per square meter in Sicily: consequently some regions have a production potential that is very high, however, it can be said that the entire national territory is characterized by very favorable conditions for the installation of plants for the production of solar power[22]. Italy's total electricity consumption was 302.75 terawatt-hour (TWh) in 2020, of which 270.55 TWh (89.3%) was produced domestically and the remaining 10.7% was imported [23].

1.4.1 Solar Power Production in Italy: Solar power accounted for 7% of the electricity generated in Italy during 2013, ranking first in the world. By 2017, that number was close to 8%, which was beaten only by Germany in Europe [24], with more than 730 000 solar power plants installed in Italy and a total capacity of 19.7 GW [24]. In 2018 capacity exceeded the 20 GW milestone and The "National

Energy Strategy", SEN, published in 2017 outlined the ambition to reach 50 GW by 2030 [25]. Sun energy currently produces around 26% of all renewable energy in the country. The 15 MWt Archimede solar field is a thermal field at Priolo Gargallo near Syracuse. The plant was inaugurated on 14 July 2010 [26,27&28] and continues to be operational in a solar field of 31,860 square meters [29]. It is the first concentrated solar power plant to use molten salt for heat transfer and storage which is integrated with a combined-cycle gas facility [26,28,30&,31]. Upon generating thermal energy, two tanks are available to store thermal energy for up to 8 hours [32]. The two other CSP systems are the ASE demo plant [33], which uses parabolic trough technology to focus solar energy, and the Rende-CSP plant, which uses Linear Fresnel reflector technology to focus solar energy to one point of fluidized storage consisting of oil [34].

Daniele Cocco et. al [35] discussed a hybrid CSP–CPV system for improving the dispatchability of solar power plants. The integrated management strategy provides the constant output curve.

1.5 China

China is a country located in East Asia with an area of 9,596,960 km^2 (3,705,410 sq mi) [36]. Solar energy, as a clean energy source and one kind of renewable energy, is abundant in China. More than two-thirds of area in China receives an annual solar radiation that **exceeds 5.9 GJ/m^2** with more than 2200 h sunshine per annum [37]. China energy consumption data was reported at 7,511.000 kWh bn in 2020. This records an increase from the previous number of 7,486.610 kWh bn for 2019. China Energy Consumption data is updated yearly, averaging 1,347.238 kWh bn from Dec 1980 to 2020, with 41 observations. The data reached an all-time high of 7,511.000 kWh bn in 2020 and a record low of 300.630 kWh bn in 1980 [38].

1.5.1 Solar Power Production in China:

At the end of 2020, China's total installed photovoltaic capacity was 253 GW, accounting for one-third of the world's total installed photovoltaic capacity (760.4 GW) [39]. China has large potential for concentrated solar power (CSP), especially in the south-western part of the country [40]. Many plants planned or under construction [41] such as: 1 MW Badaling Pilot Project — collaboration between the Institute of Electrical Engineering (IEE) and the Chinese Academy of Sciences (CAS), 100 MW project in Sichuan Abazhou by Tianwei New Energy (Aba) and 50 MW (TBD) by China Huadian Corporation etc.

Xingang Zhao et. al [42] discussed the distributed solar photovoltaics in China: Policies and economic performance. The result show that good economic benefits has been achieved by distributed PV systems with high generating efficiency. Zhao Zhu et. al [43] discussed the electricity generation costs of concentrated solar power technologies in China based on operational plants. They concluded that over other CSP technology variants, the utility scale tower CSP plants in China is cost advantageous. Hanfang Li et. al [44] discussed the research on the policy route of China's distributed photovoltaic power generation. To support the development of distributed photovoltaics, the innovative business model and financial support model are the important measure by the year 2025. Fair market competition environment can be created by the reduction in the price gap between power generation and distributed energy generation (non fossil energy generation), it also reduces government subsidies. Muhammad Awais Gulzar et. al [45] discussed the China's Pathway towards Solar Energy Utilization: Transition to a Low-Carbon Economy. Solar energy has low cost and low emission, so the solar energy produces low-carbon economy. China's solar water heating industry cost benefit's analysis indicates that solar energy not only provides economic benefit to the society but also provides environmental benefits to the earth.

1.6 Japan

The territory covers 377,976.41 km² (145,937.51 sq mi) [46]. It is the fourth largest island country in the world and the largest island country in East Asia[47].Japan has an insolation of about 4.3 to 4.8 kWh/(m²·day) [48]. Japan energy consumption per capita decreased from 4.1 toe in 2000 to 3.1 toe in 2020. Electricity consumption per capita was around 7200 kWh in 2020 [49].

1.6.1 Solar Power Production in Japan: Solar power has become an important national priority since the country's shift in policies toward renewable energy after the Fukushima Daiichi nuclear disaster in 2011[50,51]. Japan was the world's second largest market for solar PV growth in 2013 and 2014, adding a record 6.97 GW and 9.74 GW of nominal nameplate capacity, respectively. By the end of 2017, cumulative capacity reached 50 GW, the world's second largest solar PV installed capacity, behind China [52,53].Overall installed capacity in 2016 was estimated to be sufficient to supply almost 5% of the nation's annual electricity demand [52].The government set solar PV targets in 2004 and revised them in 2009 28 GW W of solar PV capacity by 2020, 53 GW of solar PV capacity by 2030&10% of total domestic primary energy demand met with solar PV by 2050 [54].

Qimei Chen et. al [55] discussed the Knowledge Mapping of Concentrating Solar Power Development Based on Literature Analysis Technology. This paper summarize topic clustering and bibliometric analysis of papers and patents in the CSP field. CSP technology has great impact for the sustainable development of human society. This paper summarizes the discipline distribution of CSP research papers (such as energy and fuels, thermodynamics, engineering chemicals and optics etc.) and also the countries/ region with their no. of patents. Japan has total416 patent in CSP technology till 2020, China, USA, Germany and India has 1559,576,343 and 180 patents respectively. The development and competiveness in the field of CSP technology increases day by day. Veronica Bermudez [56] discussed the Japan, the new "El Dorado" of solar PV? Japan increases its electricity production (using PV) to 12% by 2030,in 2018 it was only 4.5%.

1.7 Vietnam

Vietnam is located on the eastern margin of the Indochinese peninsula and occupies about 331,211.6 square kilometers [57].According to a recent mapping project by a Spanish research consortium estimations of overall solar resources in Viet Nam show an average GHI of 4-5 kWh/m²/day in most regions of southern, central and partially even northern Viet Nam (corresponding to 1,460-1,825 Wh/m²/year) and peak irradiation levels of up to 5.5 kWh/m²/day on average in some southern regions (corresponding to about 2,000 kWh/m²/year) [58].Vietnam demand is expected to increase from 265-278 TWh in 2020 to 572-632 TWh in 2030. To meet the growing demand, Vietnam needs 60,000MW of electricity by 2020, 96,500MW by 2025, and 129,500MW by 2030 [59].

1.7.1 Solar Power Production in Vietnam: As of 2020, solar and capacity in Vietnam was 16.6 gigawatts (GW). Vietnam plans to increase solar capacity to 18.6 GW and wind capacity to 18.0 GW by 2030 [60].

Eleonora Riva Sanseverino et. al [61] discussed the review of Potential and Actual Penetration of Solar Power in Vietnam. Vietnam has excellent potential for solar power production with the average solar radiation reaching up to 5Kwh/m This paper reviews the key policies and scenarios for developing solar power in Vietnam. This paper summarize the barriers and challenges for developing the solar power in Vietnam such as institutional issues, technical issues and economic and financial issues. Government needs to continuously improve the existing policies to ensure the sustainable development of solar energy.

1.8 Mexico

Mexico covers 1,972,550 square kilometers (761,610 sq mi), making it the world's 13th-largest country by area [62]Solar power in Mexico has the potential to produce vast amounts of energy. 70% of the country has an insolation of greater than 4.5 kWh/m²/day [63]Mexico electricity consumption per capita reached around 2 100 kWh in 2020 [64].

1.8.1 Solar Power Production in Mexico: Mexico was the second largest solar generator in Latin America in 2016, with 180 MW installed capacity and more than 500 MW under construction [65]. A solar trough based 14 MW plant will use a combined cycle gas turbine of 478 MW to provide electricity to the city of Agua Prieta, Sonora. The World Bank has financed this project with US$50 million [66]. A 450 MW concentrated photovoltaics plant is planned for Baja California [67].

Julia Mundo-Hernández et.al [68] discussed an overview of solar photovoltaic energy in Mexico and Germany. This paper gives an overview of solar energy potential in Mexico, electricity cost in Mexico and regulation and financing in Mexico. Mexico is located in the so called "solar belt" with radiation exceeding 5kwh per square mt. per day [69].Gibrán S. Alemán-Nava et. al [70] discussed the renewable energy research progress in Mexico: A review. This paper summarizes the expected addiditional power generation capacity under self-supply scheme (2010-2050), By 2050 solar self supply power generation will be 601 MW.

1.9 Chile

Chile covers an area of 756,096 square kilometres (291,930 sq mi) [71].Average annual DNI of 3800 kWh/m2 is achieved only in the Atacama Desert [72]. Energy consumption per capita is around 2 toe. The country's electricity consumption per capita is around 4 000 kWh [73].

1.9.1 Solar Power Production in Chile: In 2018 Chile produced about 7% of its electricity from solar power [74]. As of yearend, it had 2137 MW of solar PV capacity [75]. In July 2020 installed solar capacity had risen to 3104 MW, with another 2801 MW under construction [76]. Northern Chile has the highest solar incidence in the world [77]. In October 2015 Chile's Ministry of Energy announced its "Roadmap to 2050: A Sustainable and Inclusive Strategy", which plans for 19% of the country's electricity to be from solar energy [78].

Gonzalo Ramírez-Sagner et. al [79] discussed that concentrating solar power (csp) plants fit perfectly with chilean mining industry. optimal design challenge. This research work focus to provide 100% of electricity demand of real mining operations by using hybris CSP-PV system.R. Menaa discussed et. al [80] The impact of concentrated solar power in electric power systems: A Chilean case study. This paper focus on the integration of CSP with thermal energy storage in electric power system. The model which was considered in this paper for twenty year planning horizon, rangigng from 2018 until 2037, determining the optimal investment on generation and transmission asset. The model was prepared to capture the hourly operational dynamics of the system by considering multiple representative days for each of its investment period.

1.10 India

India is the seventh-largest country in the world, with a total area of 3,287,263 square kilometres (1,269,219 sq mi) [81,82&83].An average DNI variation of 5.21 kWh/m² and GHI variation of 5.72 kWh/m² was observed. The precise summary statistics stated that maximum and minimum values of DNI range from 3.72 kWh/m² to 5.59 kWh/m² with an average value of 5.18 kWh/m² and likewise maximum and minimum values of GHI ranges from 4.91 kWh/m^2 to 5.99 kWh/m^2 with an average value of 5.71 kWh/m^2[84].India is the world's third largest producer and third largest consumer of electricity. The national electric grid in India has an installed capacity of 383.37 GW as of 31 May 2021. During the fiscal year (FY) 2019-20, the gross electricity generated by utilities in India was 1,383.5 TWh and the total electricity generation (utilities and non utilities) in the country was 1,598 TWh. The gross electricity consumption in FY2019 was 1,208 kWh per capita. The government's National Electricity Plan of 2018 states that the country does not need more non-renewable power plants in the utility sector until 2027, with the commissioning of 50,025 MW coal-based power plants under construction and addition of 275,000 MW total renewable power capacity after the retirement of nearly 48,000 MW old coal-fired plants. It is expected that non-fossil fuels generation contribution is likely to be around 44.7% of the total gross electricity generation by the year 2029-30 [85].

1.10.1 Solar Power Production in India: Solar power in India is a fast developing industry as part of the renewable energy in India. The country's solar installed capacity was 44.3 GW as of 31 August 2021 [86]. The Indian government had an initial target of 20 GW capacity for 2022, which was achieved four years ahead of schedule [87]. In 2015 the target was raised to 100 GW of solar capacity (including 40 GW from rooftop solar) by 2022, targeting an investment of US$100 billion [88&8]. The Ministry of New and Renewable Energy had stated that a further 36.03 GW (as of January 31, 2021) of solar projects are under various stages of implementation and 23.87 GW are in the tendering process [90]. The International Solar Alliance (ISA), proposed by India as a founder member, is headquartered in India. India has also put forward the concept of "One Sun One World One Grid" and "World Solar Bank" to harness abundant solar power on global scale [91&92]. The country's solar capacity reached 19.7 GW by the end of 2017, making it the third-largest global solar market [9]. This auxiliary power requirement can be made available from cheaper solar PV plant by envisaging hybrid solar plant with a mix of solar thermal and solar PV plants at a site. Also to optimise the cost of power, generation can be from the cheaper solar PV plant (33% generation) during the daylight whereas the rest of the time in a day is from the solar thermal storage plant (67% generation from Solar power tower and parabolic trough types) for meeting 24 hours baseload power [94].

Deepak Bishoyia et. al [95] discussed the Modeling and performance simulation of 100 MW PTC based solar thermal power plant in Udaipur India. This paper evaluated the modeling and thermal performance of 100 MW Parabolic trough solar thermal power plant with 6 hour of thermal energy storage by using System Advisor Model (SAM). The efficiency of the designed hypothetical CSP Plant is 21% with annual electricity generation of 285,288,352 Kwh.Saurabh Pathak et. [96] disussed the design Investigation of 5 kW Organic Rankine Cycle (ORC) System Using Diffusion Absorption Refrigeration (DAR) for Cooling and Power Generation for India. This paper concluded that combined ORC and DAR system to produce power and cooling both was feasible together. Cooling water temperature reduction from 23.5 C to 15.5 C leads to 3 C reduction in condensing temperature and around 7% increment in thermal efficiency.Saurabh Pathak et. al [97] summarized a review on the Performance of Organic Rankine Cycle with Different Heat Sources and Absorption Chillers. This paper summarizes that ORC is best suited cycle for waste heat recovery applications and performance of ORC with different heat source like solar, industrial waste heat, geothermal energy, biomass etc. has been discussed.This research paper concluded that integration of DAR and ORC is best for sustainable development due to having no moving parts and requires no electrical energy to operate.

2. CONCLUSION

The whole world focuses on harnessing the solar energy a step towards a green energy and carbon zero future. Such as Morocco is home to the world's biggest CSP project – the Ouarzazate Solar Power Station, which alone has a capacity of 510MW also known as the Noor Power Station [98]. The North African country aims to source 52% of its energy from renewables by 2030. Morocco is continuing efforts to diversify its energy mix and ramp up its green energy production by expanding its wind, hydraulic, and solar infrastructure [99].Concentrated solar power plants account for 13.5% of South Africa's overall solar power capacity. The country added 100MW of new CSP capacity in 2018, taking its overall installed capacity in CSP to 400MW. It, further, targets to install 1GW of CSP capacity by 2030 [98]. In 2010, the Integrated Resource Plan (IRP) was released by the South African Department of Energy (DoE). According to the IRP, 42% of the total additional new capacity until 2030 was to be sourced from renewable energy sources, including 1,000 MW of CSP [100].The Renewable Energy Project Development Office of Saudi Arabia (REPDO) published the *"Saudi Arabia's Renewable Energy Program 2030"* earlier on January 9th, the country's renewables target for 2023 has been revised up from 9.5 GW to 27.3 GW, and that for 2030 set at 58.7 GW, of which 40GW PV, 16GW Wind and 2.7GW CSP [101]. Development is underway on the Italian island of Sardinia of a 1.2 MW hybrid microgrid that incorporates concentrating solar power (CSP), a diversion from the more common use of solar photovoltaic (PV) panels in microgrids. The microgrid also includes 0.6 MW of CSP, including thermal storage, along with 0.6 MW of concentrating photovoltaics. French company Electro Power Systems (EPS) is supplying a battery-based energy storage system for the hybrid microgrid. The 0.5 MWh storage system will be used to help stabilize interimittent renewables [102]. Under the "Net Zero" goals of China CSP is getting new momentum and will play a critical role in the future energy mix featuring high percentage of REs in China and the rest of world. CSP, instead of stand alone power plant, will be an ideal "partner" to solve the curtailment of PV and Wind during the day time and continue to meet the peak load during the evening, due to the long duration of thermal energy storage system [103].A tremendous amount of untapped solar energy resource potential exists in Vietnam, however. Estimates have pegged the country's solar power potential at 60–100 GWh per year for concentrated solar power and 0.8-1.2 GWh per year for solar photovoltaic (PV) energy. Turning to Solar, Vietnam government aims to raise solar power capacity to 0.5 % of national output by 2020, 3.3 % by 2030 and 20 % by 2050. It's targeting 850 megawatts (MW) of installed solar power generation capacity by 2020, 4 gigawatts (GW) by 2025 and 12 GW by 2030 [104]. Mexico can increase the use of renewable energy in its energy mix from 4.4 per cent in 2010 to 21 per cent by 2030, according to a report released today by the International Renewable Energy Agency (IRENA). Renewable Energy Prospects: Mexico, prepared in collaboration with the Mexican Energy Secretariat (SENER), also finds that Mexico could generate up to 46 per cent of its electricity by 2030 from renewable sources including wind, solar, hydropower, geothermal and biomass – a six fold increase from today's levels [105].Chile is currently hosting a 110 MW CSP project built by Spanish renewable energy company Acciona – the Cerro Dominador plant. The project is located in the Atacama Desert, renowned as the region with the highest concentration of solar irradiation in the world [106].

India's CSP capacity accounts for a small percentage of its total installed solar capacity of 27GW. Reliance Power's 100MW Dhursar CSP plant in Pokhran tehsil in Jaisalmer district, Rajasthan, is one of the projects commissioned in recent years in the country. Built with an investment of $341m, the project is one of the world's biggest CSP projects based on compact linear fresnel reflector (CLFR) technology, which was provided Areva Solar [98].To reduce its dependence on coal-fired energy production, a total of 470 MW under CSP projects were planned in India under the phase I (2010-2013) of the ambitious Jawaharlal Nehru National Solar Mission (JNNSM). However, only 228.5 MW of total planned capacity is found operational in India as of 2018. The government of India has set up a target of 2000 MW off-grid solar PV application under its National Solar Mission that is to be achieved between 2017-2022.CSP technologies can be commissioned in the states with high solar irradiance like Rajasthan, Gujarat and Tamil Nadu and can be used as alternative energy sources instead of commissioning new fossil-fuel power plants like natural gas, lignite and coal [107].As of July 2021, India had 96.96 GW of renewable energy capacity, and represents 25.2% of the overall installed power capacity, providing a great opportunity for the expansion of green data centers. The country is targeting about 450 Gigawatt (GW) of installed renewable energy capacity by 2030 – about 280 GW (over 60%) is expected from solar [108].India, solar power plays a vital role to in the global search for nature-based and technology-driven solutions which are critical to accelerating the move towards a zero carbon future [109].Indian market continuously trying to increase the share of solar energy for sustainable development. For constant power output there is a need to work on PV and CSP hybridization, PV is used to produce electricity directly and CSP store the thermal energy at that time period and release that energy when PV is not able to produce electricity. Every day, the sun gives off far more energy than we need to power everything on earth so we should focus on to the international trading system of solar energy.

REFERENCES

[1] T. Kousksou, A. Allouhi, M. Belattar et al., "Renewable energy potential and national policy directions for sustainable development in Morocco," Renewable and Sustainable Energy Reviews, vol. 47, pp. 46–57, 2015.

[2] Y. El Mghouchi, T. Ajzoul, and A. El Bouardi, "Prediction of daily solar radiation intensity by day of the year in twenty four cities of Morocco," Renewable and Sustainable Energy Reviews, vol. 53, pp. 823–831, 2016.

[3] https://www.worldometers.info/electricity/morocco-electricity/

[4] AfDB helps fund $1.44bn Moroccan solar project

[5] Morocco Building 500 MW Solar Power Project

[6] Louis Boisgibault, Fahad Al Kabbani (2020): Energy Transition in Metropolises, Rural Areas and Deserts. Wiley - ISTE. (Energy series) ISBN 9781786304995.

[7] Loubna Bousselamt,Mohamed Cherkaoui "Modelling and Assessing the Performance of Hybrid PV-CSP Plants in Morocco: A Parametric Study" Hindawi, International Journal of Photoenergy,Article ID 5783927, 15 pages,Volume 2019.

[8] Ayat-allah Bouramdane ,Alexis Tantet,Philippe Drobinski "Adequacy of Renewable Energy Mixes with Concentrated Solar Power and Photovoltaic in Morocco:Impact of Thermal Storage and Cost" (http://creativecommons.org/licenses/by/4.0/),Volume 13 Energies 2020.

[9] L. Bousselamti,M.Cherkaoui, M.Labbadi "Study of hybrid PV-CSP plants considering two dispatch strategies in Ouarzazate" Proceedings of the 8th International Conference on Systems and Control, Marrakech, Morocco, October23-25, 2019.

[10] Sebastian-James Bode, Alberto Cuellar, Iñaki Perez, "Retrofitting Operating CSP Plants with PV to PowerAuxiliary Loads – Technical Consideration and Case Study" AIP Conference Proceedings 2126, 090003 (2019); https://doi.org/10.1063/1.5117605, July2019.

[11] https://en.wikipedia.org/wiki/Energy_in_South_Africa

[12] "Monitoring of Renewable Energy Performance 2016". ww.nersa.org.za. Retrieved 2017-06-03.

[13] "Solar energy in South Africa - REVE". www.evwind.es. Retrieved 16 April 2018.

[14] Ahmed Bilal Awan, Muhammad Zubair, R.P. Praveen, Abdul Rauf Bhatti "Design and comparative analysis of photovoltaic and parabolic trough based CSP plants" Solar Energy 183,551–565,2019.

[15] Sulaiman AlYahya, Mohammad A. Irfan "Analysis from the new solar radiation Atlas for Saudi Arabia" ScienceDirect,Solar Energy 130,116–127,2016.

[16] "Energy Information Agency, Country Analysis Briefs 2007". *Eia.doe.gov*. Retrieved 28 April 2011.

[17] http://www.eia.gov/countries/cab.cfm?fips=SA

[18] Saudi Arabia to become solar powerhouse - Nov. 21, 2011 (cnn.com)

[19] Louis Boisgibault, Fahad Al Kabbani (2020): Energy Transition in Metropolises, Rural Areas and Deserts. Wiley - ISTE. (Energy series) ISBN 9781786304995.

[20] Wael Mahdi and Marc Roca. "Saudi Arabia Plans $109 Billion Boost for Solar Power." Bloomberg News. May 11, 2012 9:23 AM CT

[21] Mario Petrollese, Daniele Cocco "Optimal design of a hybrid CSP-PV plant for achieving the full dispatchability of solar energy power plants" Solar Energy 137,477–489,2016.

[22] Photovoltaic Geographical Information System (PVGIS)-http://re.jrc.ec.europa.eu/pvgis/)

[23] https://en.wikipedia.org/wiki/Electricity_sector_in_Italy

[24] "2018 snapshot of global photovoltaic markets. Photovoltaic power systems programme. Report IEA PVPS T1-33:2018".

[25] "PVPS Annual Report 2018"

[26] Backwell, Ben (2010-07-14). "Enel starts up its Archimede plant in world first for CSP". ReCharge. NHST Media Group. Retrieved 2010-07-15.

[27] Babington, Deepa (2010-07-14). "Sicily plant offers Italy new impetus on solar front". Reuters. Retrieved 2010-07-15.

[28] "At Priolo Enel inaugurates the "Archimede" power plant" (Press release). Enel. 2010-07-14. Retrieved 2010-07-15.

[29] Kabir, Ehsanul; Kumar, Pawan; Kumar, Sandeep; Adelodun, Adedeji A.; Kim, Ki-Hyun (2018). "Solar energy: Potential and future prospects". Renewable and Sustainable Energy Reviews. 82:894900. doi:10.1016/j.rser.2017.09.094. ISSN 1364-0321.

[30] "ENEL opens "world's first" molten-salt/solar plant". The Engineer. Centaur Media plc. 2010-07-14. Retrieved 2010-07-15.

[31] Popham, Peter (2007-03-28). "Sicily to build world's first solar power plant". The Independent. Retrieved 2010-07-15.

[32] "Concentrating Solar Power Projects - Archimede | Concentrating Solar Power | NREL". www.nrel.gov. Retrieved 2018-04-06.

[33] "Concentrating Solar Power Projects - ASE Demo Plant | Concentrating Solar Power | NREL". www.nrel.gov. Retrieved 2018-04-06.

[34] "Concentrating Solar Power Projects - Rende-CSP Plant | Concentrating Solar Power | NREL". www.nrel.gov. Retrieved 2018-04-06.

[35] Daniele Cocco, Luca Migliari, Mario Petrollese, "A hybrid CSP–CPV system for improving the dispatchability of solar power plants" Energy Conservation & Management, 312-323, 2016.

[36] The World Factbook. CIA. Retrieved 2015-12-31.

[37] Li J, Wan Y-H, Ohi JM. "Renewable energy development in China: resource assessment, technology status, and greenhouse gas mitigation potential."Applied Energy 1997;56:381–94

[38] https://www.ceicdata.com/en/china/electricity-consumption/energy-consumption-electricity

[39] "Snapshot 2021". IEA-PVPS. Archived from the original on 27 April 2021. Retrieved 6 June 2021.

[40] Wang, Jun; Yang, Song; Jiang, Chuan; Zhang, Yaoming; Lund, Peter D. (April 2017). "Status and future strategies for Concentrating Solar Power in China". Energy Science & Engineering. 5 (2): 100–109. doi:10.1002/ese3.154.

[41] "China to Have 3 GW of Concentrated Solar Thermal Power (CSP) by 2020 - CleanTechnica". cleantechnica.com. 30 March 2012. Archived from the original on 25 December 2014. Retrieved 20 July 2012.

[42] Xingang Zhao, Yiping Zeng, Di Zhao "Distributed solar photovoltaics in China: Policies and economic performance" Energy, 1-12, 2015.

[43] Zhao Zhu, Da Zhang, Peggy Mischke, Xiliang Zhang "Electricity generation costs of concentrated solar power technologies in China based on operational plants" Energy,1-10, 2015.

[44] Hanfang Li , Hongyu Lin , Qingkun Tan, Peng Wu, Chengjie Wang , Gejirifu D , Liling Huang "Research on the policy route of China's distributed photovoltaic power generation" Energy Reports 6, 254-263, 2020

[45] Muhammad Awais Gulzar, Haroon Asghar Jinsoo Hwang and Waseem Hassan "China's Pathway towards Solar Energy Utilization: Transition to a Low-Carbon Economy" international jouran of Environment Research and Public Health, 2020.

[46] Reiwa 2nd year National area of each prefecture municipality (as of October 1)" (in Japanese). Geospatial Information Authority of Japan. 25 December 2020. Archived from the original on January 1, 2021. Retrieved 3 January 2021.

[47] "Island Countries Of The World". WorldAtlas.com. Archived from the original on 2017-12-07. Retrieved 2019-08-10.

[48] https://en.wikipedia.org/wiki/Solar_power_in_Japan

[49] https://www.enerdata.net/estore/energy-market/japan/

[50]   "Solar Energy in Japan – Summary". GENI. Retrieved 7 May 2012.

[51]   Chisaki Watanabe (August 26, 2011). "Japan Spurs Solar, Wind Energy With Subsidies, in Shift From Nuclear Power". Bloomberg.

[52]   Snapshot of Global Photovoltaic Markets 2017" (PDF). report. International Energy Agency. 19 April 2017. Retrieved 27 April 2017.

[53]   Pv-magazine FEBRUARY 15, 2018. "Japan will likely install 6 GW to 7.5 GW (DC) of solar in 2018, from about 7 GW in 2017..."

[54]   Yamamoto, Masamichi; Ikki, Osamu (2010-05-28). "National survey report of PV Power Applications in Japan 2009" (PDF). International Energy Agency. Retrieved 2017-04-02.

[55]   Qimei Chen, Yan Wang, Jianhan Zhang and Zhifeng Wang "The Knowledge Mapping of Concentrating Solar Power Development Based on Literature Analysis Technology" Energies 2020, 13, 1988; doi:10.3390/en13081988.

[56]   Veronica Bermudez "Japan, the new "El Dorado" of solar PV?" Journal of Renewable and Sustainable Energy 10, 020401 (2018); https://doi.org/10.1063/1.5024431.

[57]   https://en.wikipedia.org/wiki/Geography_of_Vietnam

[58]   CIEMAT et al. (2015). Maps of Solar Resource and Potential in Vietnam. Ha Noi: CIEMAT, CENER & IDAE with support from AECER in collaboration with GDE/MoIT. Published: http://bit.ly/1Q0FEhb

[59]   https://www.vietnam-briefing.com/news/vietnams-push-for-renewable-energy.html/

[60]   https://www.eia.gov/todayinenergy/detail.php?id=48176

[61]   Eleonora Riva Sanseverino , Hang Le Thi Thuy , Manh-Hai Pham, Maria Luisa Di Silvestre , Ninh Nguyen Quang and Salvatore Favuzza , "Review of Potential and Actual Penetration of Solar Power in Vietnam" Energies 2020, 13, 2529; doi:10.3390/en13102529.

[62]   Mexico. *The World Factbook*. Central Intelligence Agency.

[63]   https://en.wikipedia.org/wiki/Solar_power_in_Mexico

[64]   https://www.enerdata.net/estore/energy-market/mexico/

[65]   Mike, Munsell (June 14, 2016). "3 Fast Facts About Latin America's Solar Market".

[66]   "Cumulative and Newly-Installed Solar Photovoltaics Capacity in Ten Leading Countries and the World, 2009". Earth Policy Institute. 2010-09-21. Retrieved 2010-09-22.

[67]   First 50 Megawatts of Large Solar Power Plant in Baja California Archived 2012-08-23 at the Wayback Machine

[68]   Julia Mundo-Hernández , Benito de Celis Alonso, Julia Hernández-Álvarez, Benito de Celis-Carrillo "An overview of solar photovoltaic energy in Mexico and Germany" Renewable and Sustainable Energy Reviews 31 (2014) 639-649.

[69]   EPIA. Unlocking the sunbelt potential of photovoltaics. 2010.

[70]   Gibrán S. Alemán-Nava , Victor H. Casiano-Flores , Diana L. Cárdenas-Chávez, Rocío Díaz-Chavez, Nicolae Scarlat , Jürgen Mahlknecht , Jean-Francois Dallemand, Roberto Parra "Renewable energy research progress in Mexico: A review"Renewable and Sustainable Energy Reviews 32 (2014) 140–153

[71]   "RESULTADOS CENSO 2017" (PDF). RESULTADOS DEFINITIVOS CENSO 2017. National Statistics Institute. 1 January 2018. Retrieved 18 January 2017.

[72]   Carlos Portillo, Elisa Alonso, Angel Fernández, Pablo Ferrada, Alessandro Gallo , Martin Guillaume, Aitor Marzo and Edward Fuentealba "Progress in solar energy r&d in north of chile: solar platform of atacama desert project and ongoing activities" Solar Word Congress 2015 Daegu ,Korea, 08 – 12 November 2015.

[73]   https://www.enerdata.net/estore/energy-market/chile/

[74]   Renewable Capacity Statistics (PDF). IRENA. 2019. pp. 24–26. ISBN 978-92-9260-123-2. Retrieved 3 May 2019.

[75]   "IEA PVPS Snapshot of Global PV 2019" (PDF). IEA.

[76]   "La solar alcanzó en julio una capacidad instalada de 3.104 MW en Chile". 4 September 2020

[77]   Inter-American Development Bank (15 December 2011). "Renewable energy to power irrigation in the Atacama desert". Retrieved 22 July 2014.

[78]   Chile sets 70 pct. non-conventional renewable energy target for 2050, Fox News Latino from EFE, 30 September 2015

[79]   Gonzalo Ramírez-Sagne, Frank Dinter and Mercedes Ibarra "Concentrating Solar Power (Csp) Plants Fit Perfectly With Chilean Mining Industry. Optimal Design Challenge" IEA SHC International Conference on Solar Heating and Cooling for Buildings and Industry 2019

[80]   R. Menaa, R. Escobar , Á. Lorca, M. Negrete-Pinceticc, D. Olivaresc, "The impact of concentrated solar power in electric power systems: A Chilean case study" Applied Energy 235 (2019) 258–283

[81]   "India". Encyclopædia Britannica. Retrieved 17 July 2012. Total area excludes disputed territories not under Indian control.

[82]   "India at a Glance: Area". Ministry of Home Affairs: Government of India. 2001. Retrieved 9 September 2020.

[83]   "Jammu and Kashmir - CIA" (PDF). Central Intelligence Agency. 2002. Retrieved 9 September 2020.

[84]   DeepakKumar Satellite-based solar energy potential analysis for southern states of India **Energy Reports** Volume 6, November 2020, Pages 1487-150

[85]   https://en.wikipedia.org/wiki/Electricity_sector_in_India

[86]   "Physical Progress (Achievements)". Ministry of New & Renewable Energy. Retrieved 18 July 2020.

[87]   "India hits 20 GW solar capacity milestone". The Times of India. 31 January 2018. Retrieved 4 February 2018.

[88]   Krishna N. Das (2 January 2015). "India's Modi raises solar investment target to $100 bln by 2022". Reuters. Retrieved 2 January 2015.

[89]   Kenning, Tom (2 July 2015). "India releases state targets for 40GW rooftop solar by 2022". PV Tech. Retrieved 29 July 2016.

[90]   "List of solar parks in India". mnre.gov.in (MS Word Document). Archived from the original on 11 December 2019. Retrieved 7 September 2019.

[91]   Tayal, Manu (27 May 2020). "MNRE Invites Proposals to Develop Institutional Framework for 'One Sun One World One Grid' Implementation". Saur Energy. Retrieved 31 May2020.

[92]   ^ "India Set To Propose World Solar Bank & Mobilize $50 Billion In Solar Funding". Clean Technica. 26 July 2020. Retrieved 27 July 2020.

[93]   Prateek, Saumy (14 March 2018). "Four States Accounted for 75 Percent of India's Utility-Scale Solar Installations in 2017". Mercom India. Retrieved 20 June 2021.

[94]   Kraemer, Susan (25 August 2015). "Cheap Baseload Solar At Copiapó Gets OK In Chile". Clean Technica. Retrieved 1 September 2017.

[95]   Deepak Bishoyia, K. Sudhakara, "Modeling and performance simulation of 100 MW PTC based solar thermal power plant in Udaipur India" Case Studies in Thermal Engineering 10 (2017) 216–226

[96]   Saurabh Pathak and S.K. Shukla Design "Investigation of 5 kW Organic Rankine Cycle (ORC) System Using Diffusion Absorption Refrigeration (DAR) for Cooling and Power Generation for India" Asian Journal of Water, Environment and Pollution, Vol. 16, No. 2 (2019), pp. 35–42. DOI 10.3233/AJW190017

[97]   Saurabh Pathak & S.K. Shukla "A Review on the Performance of Organic Rankine Cycle with Different Heat Sources and Absorption Chillers Distributed Generation & Alternative" Energy Journal ISSN: 2156-3306 (Print) 2156-6550 (Online) Vol. 33, No.2   2018.

[98]   https://www.nsenergybusiness.com/features/concentrated-solar-power-countries/

[99]   https://heliosesp.com/moroccos-noor-concentrated-solar-power-projects-support-africas-energy-transition/

[100]   https://www.solarpaces.org/csp-technologies/csp-potential-solar-thermal-energy-by-member-nation/south-africa/

[101]   https://heliosesp.com/saudi-arabia-targets-2-7gw-concentrated-solar-power-in-2030/

[102]   https://heliosesp.com/hybrid-microgrid-in-italy-to-include-concentrated-solar-power/

[103]   https://heliosesp.com/concentrated-solar-power-opportunities-under-chinas-net-zero-ambition/

[104]   https://solarmagazine.com/solar-profiles/vietnam/

[105]   https://heliosesp.com/mexico-can-add-60-gw-of-wind-power-and-solar-energy-by-2030-irena/

[106]   https://www.pv-magazine.com/2021/09/01/world-record-low-bid-of-0-0399-kwh-for-csp-technology-in-chiles-renewables-auction\

[107]   https://www.cag.org.in/newsletters/public-newsense/concentrated-solar-power-csp-potential-solution-india

[108]   https://www.ibef.org/industry/renewable-energy.aspx

[109]   https://www.business-standard.com/article/economy-policy/india-s-solar-power-efforts-an-example-to-world-says-prince-charles-121063001855_1.html

9

Renewable Energy the Clean Facts

Manoj Kumar Singh[1*] and Bharat Raj Singh[2]
[1*]Department of Mechanical Engineering,
Babu Sunder Singh Institute of Tech and Management, Lucknow, UP, India
e-mail: mnjsingh567@gmail.com
[2]School of Management Sciences, Lucknow
e-mail: brsinghlko@yahoo.com

ABSTRACT

Renewable energy resources are sustainable, a few are not. as an instance, a few biomass sources are taken into consideration unsustainable at modern charges of exploitation. Renewable energy is regularly used for electricity era, heating and cooling. Renewable electricity projects are usually huge-scale, however they may be additionally acceptable to rural and remote regions and growing international locations, where energy is frequently critical in human improvement. Renewable electricity is often deployed collectively with in addition electrification, which has several advantages: electricity can circulate heat or objects effectively, and is easy at the factor of consumption. Further, electrification with renewable electricity is greener and therefore results in tremendous reductions in primary strength necessities.

From 2011 to 2021, renewable power has grown from 20% to 28% of world power supply. Use of fossil electricity shrank from 68% to 62%, and nuclear from 12% to 10% the share of hydropower reduced from 16% to 15% at the same time as electricity from sun and wind increased from 2% to 10%. Biomass and geothermal power grew from 2% to 3%. There are 3,146 gigawatts hooked up in one hundred thirty five nations, while 156 countries have legal guidelines regulating the renewable electricity area. In 2021, China accounted for nearly half of the global growth in renewable strength.

Keywords: *Renewable energy, Biomass sources, Geothermal power, Biomass, Solar energy, Wind energy and Hydropower.*

1. INTRODUCTION

Renewable energy resources (RES) that use indigenous sources have the potential to provide electricity with negligible emissions of air pollutants and inexperienced residence gases [1]. Renewable electricity technologies produce marketable strength by way of changing natural phenomena/sources into useful energies. The usage of renewable strength sources is a promising prospect for the future as an alternative to traditional power. Therefore, an try has been made thru this paper to check the availability of renewable energy options in India, and presents information about the present day reputation of renewable, future potentials of their makes use of, primary achievements, and current authorities guidelines, delivery and outreach in Indian context. It paints a outstanding common picture of renewable energy assets and position of India on worldwide map in using these resource the world electricity forum has predicted that fossil-based totally oil, coal and gas reserves will be exhausted in less than any other 10 a long time. Fossil fuels account for over 79% of the primary energy ate up inside the world, and 57.7% of that amount is used in the transport sector and are diminishing hastily [2]. The exhaustion of herbal assets and the elevated call for of conventional strength have compelled planners and policy makers to search for trade sources. Renewable energy is electricity derived from sources which can be regenerative, and do no longer dissipate through the years. Renewable energy offers our planet a danger to lessen carbon emissions, easy the air, and positioned our civilization on a greater sustainable footing. It additionally offers countries around the arena the chance to improve their electricity safety and spur financial development. Cutting-edge biomass encompasses quite a number products derived from photosynthesis and is basically chemical solar electricity garage. Renewable power supplies 18% of the sector's final electricity intake (Figure 1), counting conventional biomass, big hydropower, and "new" renewable (small hydro, cutting-edge biomass, wind, sun, geothermal, and biofuels). Conventional biomass, normally for cooking and heating, represents about 13% and is developing slowly in some areas as biomass is used greater efficaciously or replaced by means of extra contemporary power bureaucracy. Huge hydropower represents three% and is growing modestly, often in growing nations [3]. New Renewable represents 2.Four% and are developing very hastily in developed countries and in some developing nations. Worldwide renewable energy capacity grew at costs of 15–30% yearly for many technology at some stage in the 5-12 months duration 2002–2006, together with wind power, sun warm water, geothermal heating, and rancid-grid solar PV (Figure 2) [4]. Renewable power markets grew robustly in 2008. Amongst new renewable (apart from massive hydropower), wind power became the biggest addition to renewable power ability. An envisioned $120 billion was invested in Renewable energy international in 2008, together with new ability (asset finance and tasks) and biofuels refineries (Figure 3) [5].

2. CHALLENGES AFFECTING RENEWABLE ENERGY THE CLEAN FACTS

Renewable energy resources should turn out to be the foremost energy alternative for low-carbon energy savings. Different adjustments to all power systems are needed to make renewable resources of electricity available on a massive scale. Organizing the transition of power from non-sustainable strength to renewable energy is often described as the important mission of the primary 1/2 of the 21st century.[6] .

The following are policy pointers from the take a look at that could assist mitigate weather alternate and its impact:

- All sectors and areas have the ability to make contributions by means of making an investment in renewable energy technologies and regulations to lessen them.
- Decreasing our carbon footprint thru life-style modifications and behavioural patterns can greatly make a contribution to mitigating weather change.
- Research on innovations and technologies that may lessen land use and can also reduce accidents from renewable electricity sources and the danger of resource opposition, for example in the discipline of bio-energy where food for intake competes with energy manufacturing.
- Increase international cooperation and help growing countries to expand infrastructure and modernize era for modern-day supply and sustainable power services as a method of mitigating weather trade and its effect.

3. RENEWABLE ENERGY THE CLEAN FACTS IN INDIA

India's population of greater than 1028 million is developing at an annual price of 1.58%. As fossil gas power becomes scarcer, India will face strength shortages significantly due to boom in electricity costs and electricity lack of confidence within the next few decades.

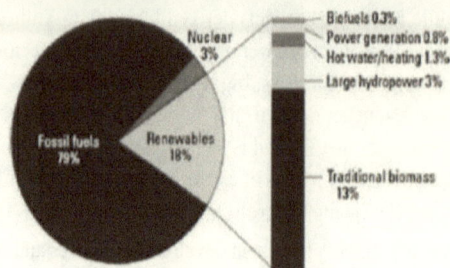

Figure 1: Renewable energy share of global final energy consumption

Expanded use of fossil fuels also reasons environmental issues each domestically and globally. The economic system of India, measured in USD trade-rate terms, is the 12th biggest in the international, with a GDP of around $1 trillion (2008). GDP increase price of 90% for the monetary year 2007–2008 which makes it the second quickest big rising economic system, after China, within the international. There is a totally excessive demand for energy, which is currently satisfied mainly by coal, foreign oil and petroleum, which apart from being a non-renewable, and therefore non-permanent solution to the energy crisis, it is also detrimental to the environment. Thus, it is imperative that India obtains energy security without affecting the booming economy, which would mean that the country must switch from the nonrenewable energy (crude oil and coal) to renewable energy. Electricity, which is currently happy especially via coal, foreign oil and petroleum, which apart from being a non-renewable, and therefore non-everlasting technique to the energy disaster, it is also unfavorable to the surroundings. Accordingly, it is imperative that India obtains power safety without affecting the booming economic system, which would imply that the United States of America must switch from the nonrenewable electricity (crude oil and coal) to renewable energy.

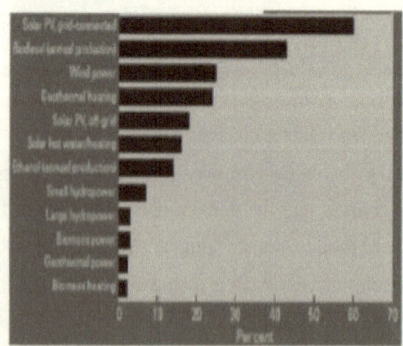

Figure 2: Average annual growth rates of renewable energy capacity, 2002–2006.

For these reasons the development and use of RES & technology are becoming crucial for sustainable monetary development of India. Expert consultation at the Asia power vision 2020, organized underneath the world power Council agreed on power call for projection [7]. The professional Committee on incorporated strength coverage in its record (IEPR2006) has estimated that by 2032, i.e., 25 years from now primary commercial electricity requirement in the United States of America might need to go up 4–five times the contemporary degree, strength technology hooked up potential 5.6–7 times the contemporary level and oil requirement by three–6 instances the modern-day stage.

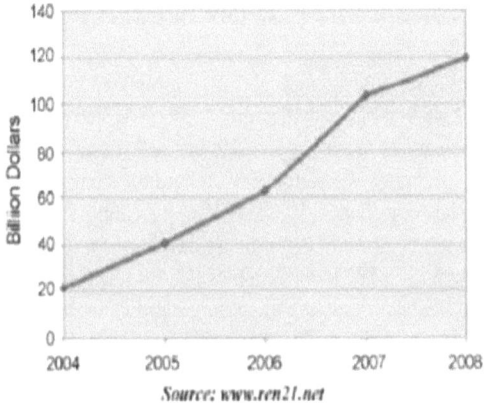

Source: www.ren21.net

Figure 3: Global investment in renewable energy, 2004-2008.

Strength is a simple requirement for monetary improvement and in each area of Indian economic system.

It is for this reason necessary that India quick appearance in the direction of new and rising renewable energy and power efficient technology in addition to put in force electricity conservation legal guidelines. Against this history, the United States urgently needs to broaden a sustainable path of power improvement. Promotion of electricity conservation and elevated use of renewable energy assets are the dual planks of a sustainable strength supply.

Thankfully, India is blessed with a ramification of renewable energy sources, like biomass, the sun, wind, geothermal and small hydropower and implementing one of the international's biggest programs in renewable power. India is decided to turning into one of the globe's main smooth power producers. The authorities of India has already made numerous provisions, and installed many companies that will help it to attain its goal. Renewable energy, apart from big hydro tasks already account for 9% of the whole hooked up electricity capacity, equal to 12,610MWof electricity. In mixture with massive hydro, the potential is extra than 34%, ie., 48,643 MW, in a complete mounted capability of a hundred and 44,980 MW. Figure4 is showing established strength potential (MW) in India.

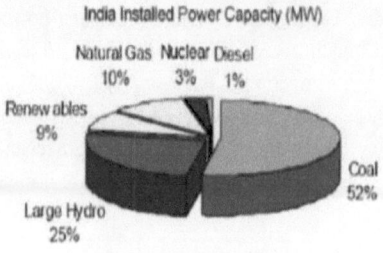

Source: CEA, 2008, MNRE, 2008

Figure 4: Installed power capacity (MW) in India as of June 2008.

The US has an expected renewable strength ability of around 85,000MW from commercially exploitable assets, i.e., wind, 45,000 MW; small hydro, 15,000MW and biomass/bioenergy, 25,000 MW. Further, India has the capability to generate 35MW in step with square kilometer the use of sun photovoltaic and solar thermal energy. Via March 2007, renewable strength, apart from hydro above 25MW hooked up ability, has contributed 10,243MW representing 7.7% of general electricity established potential. There has been out of the ordinary development in wind electricity and, with an installed capacity of over 8757 MW, India occupies the 5th function globally [8]. The role of recent and renewable power has been assuming increasing significance nowadays with the growing problem for the US of as energy security. The renewable energy enterprise has about USD 500 million as turnover, the funding being approximately USD 3 billion. Of the envisioned potential of a 100,000MW from RE handiest about 3500MW has been exploited to-date. The Indian authorities has been at work, creating a comprehensive coverage for compulsory use of renewable strength resources through biomass, hydropower, wind, solar and municipal waste inside the U.S.A. particularly for Business establishments, as well as government institutions. The predominant contribution to renewable strength funding comes from personal region participation. That is due to the aid from the authorities, which leverages the personal funding. The economic allocation for renewable energy resources vis-à-vis total allocation, but stays in the variety of 0.1% for the duration of 10th Plan length. [9].

According to the eleventh new and renewable power 5-year plan proposed through the government of India, from 2008 to 2012 the renewable energy marketplace in India will reach an predicted US $19 billion. Investments people $15 billion could be required so that you can add the approximately 15,000 megawatts (MW) of renewable strength to the existing hooked up capability. The Indian authorities has additionally set specific goals for renewable energy with the aid of 2012 it expects renewable strength to make a contribution 10% of overall electricity technology capacity and have a 4-5% percentage within the power blend. This implies that growth in renewable strength will occur at a miles faster pace than traditional energy generation, with renewable making up 20% of the 70,000MW of general additional strength deliberate from 2008 to 2012. From 2002 to 2007, there has been 3075MW of renewable grid-tied strength deliberate, but the actual ability addition passed 6000MW by way of 2006. A large percentage of this was the end result of remarkable growth of wind power in India. Wind electricity is expected to feature extra than 10,000MW of additional capacity by 2012, followed by way of small hydro (1400 MW), cogeneration (1200 MW) and biomass (500

MW). Ministry of Nonconventional power resources is focused on nation-huge resource assessment, putting in place of commercial initiatives, protection and modernization, development and up-gradation of water generators and enterprise primarily based research and development. The Ministry of recent and Renewable energy has recognized renewable strength R&D as an critical component for growing this sector. R&D subsidy is one hundred% of a challenge's cost in government R&D establishments, and 50% inside the personal zone. The R&D subsidy for the Personal area can be stronger for preliminary levels of technologies that have longer time-horizons. Renewable resources already contribute to approximately 5% of the overall strength generating ability in the USA. Over the last a long time, several renewable energy technologies had been deployed in rural and urban areas. [10].

3.1. Biomass

In latest years, the interest in the usage of biomass as an energy supply has improved and it represents about 14% of worldwide very last electricity intake [11]. Estimates have indicated that 15–50% of the sector's primary electricity use may want to come from biomass by using the 12 months 2050. Many nations have protected the elevated use of renewable resources on their political time table. Biomass is one such useful resource that might play a large position in a greater various and Sustainable energy mix. The energy obtained from biomass is a shape of renewable strength and, in principle, making use of this strength does not add carbon dioxide, a major greenhouse gasoline, to the Ecosystem, in contrast to fossil fuels. As in step with an estimate, globally photosynthesis produces 220 billion dry tonnes of biomass every 12 months with 1% conversion performance [12-14]. A biomass useful resource appropriate for power production covers a wide variety of materials, from firewood accrued in farmlands and herbal woods to agricultural and forestry vegetation grown specifically for strength manufacturing purposes. Electricity manufacturing from meals wastes or food processing wastes, especially from waste fit for human consumption oils, appears to be appealing based totally on bio-useful resource sustainability, environmental protection and monetary attention. India may be very rich in biomass and has a ability of 16,881MW (agro-residues and plantations), 5000MW (bagasse cogeneration) and 2700MW (electricity healing from waste) [8]. Biomass power generation in India is an enterprise that attracts investments of over Rs. 600 crores each 12 months, generating more than 5000 million gadgets of energy and every year employment of more than 10 million man-days inside the rural areas.

3.2. Hydropower

Hydropower is some other source of renewable power that converts the ability strength or kinetic electricity of water into mechanical power within the form of watermills, fabric machines,and so forth., or as electrical power (i.E., hydroelectricity technology). It refers back to the energy made out of water (rainfall flowing into rivers, and so forth.). Hydropower is the most important renewable energy aid getting used for the generation of strength. Only approximately 17% of the big hydel capacity of 150,000MW has been tapped up to now. Nations like Norway, Canada, and Brazil have all been utilizing more than 30% in their hydro-potential, while alternatively India and China have lagged a long way in the back of. India ranks fifth in phrases of exploitable hydro-potential within the world. In keeping with CEA (relevant power Authority), India is endowed with economically exploitable hydropower capability to the tune of 148,700 MW [15]. The dominant annual rainfall is located on the North-jap a part of India: Arunachal Pradesh, Assam, Nagaland, Manipur and Mizoram, and additionally at the west coast among Mumbai (Bombay) and Mahe. Primary hydroelectric electricity flora are positioned in Bihar, Punjab, Uttaranchal, Karnataka, Uttar Pradesh, Sikkim, Jammu & Kashmir, Gujarat, and Andhra Pradesh. In India, hydropower initiatives with a station capacity of up to twenty-five megawatt (MW) fall below the class of small hydropower (SHP). India has an predicted SHP potential of about 15,000 MW, of which about eleven% has been tapped thus far. The Ministry of New and Renewable energy (MNRE) helps SHP mission improvement at some stage in the United States of America. To this point, 523 SHP tasks with an combination installed capacity of 1705MW had been hooked up. Except those, 205 SHP projects with an combination capability of 479MW are under implementation. With a capacity addition, on a mean, of 100MW according to year and gradual lower in gestation intervals and capital costs, the SHP quarter is turning into an increasing number of competitive with other alternatives.

3.3. Wind Energy

Winds are generated with the aid of complex mechanisms regarding the rotation of the earth, heat strength from the solar, the cooling results of the oceans and polar ice caps, temperature gradients between Land and sea and the physical consequences of mountains and other obstacles. Wind is an extensively disbursed strength aid. Overall world wind potential on the cease of 2006 turned into round 72,000 MW. Wind strength is being advanced inside the industrialized international for environmental motives and it has attractions inside the growing global as it could be mounted speedy in regions wherein energy is urgently needed. Commonly it can be a price-powerful solution if fossil fuel sources are not effortlessly to be had. Further there are numerous packages for wind power in remote regions, Worldwide, either for supplementing diesel strength (which has a tendency to be high priced) or for offering farms, houses and other installations on an man or woman basis. The provision of wind varies for specific areas. Wind resources may be exploited in particular in regions wherein wind power density is at least 400W/m2 at 30 m above the floor.

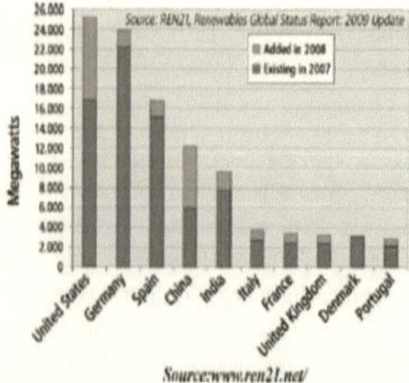

Source:www.ren21.net/

Figure 5: Wind power capacity, top ten countries, 2008

The Wind aid evaluation application is being carried out with the aid of C-wet (Centre for Wind energy era) in coordination with country nodal companies. An annual suggest wind strength density more than 200 W/m2 (watts according to rectangular meter) at 50-m height has been recorded at 211 wind tracking stations, covering thirteen states and union territories, particularly Andaman and Nicobar Islands, Andhra Pradesh, Gujarat, Karnataka, Kerala, Lakshadweep, Madhya Pradesh, Maharashtra, Orissa, Rajasthan, Tamil Nadu, Uttaranchal, and West Bengal. India's wind energy capacity has been assessed at 45,000 MW. A capacity of 8757MW up to 31 March 2008 has so far been added through wind (Figure5) [5]. The Wind energy program in India was initiated closer to the give up of the 6ᵗʰ Plan, in 1983–1984. This system objective at survey and assessment of wind sources, are setting up demonstration projects, and provision of incentives to make wind strength competitive. India is exceeded simplest by way of Germany as one of the international's fastest developing markets for wind energy. Through the mid 1990s, the subcontinent become putting in more wind generating ability than North America, Denmark, Britain, and the Netherlands. The 10 machines near Okha inside the province of Gujarat had been a number of the primary wind mills established in India. Those 15-m Vestas wind turbines forget the Arabian Sea. Now, in 2008, there may be an mounted capacity of 5310 MW; however, ten times that potential, or 45,000 MW, exists. One of a kind types of Wind energy mills utilized in India for Off grip power generation, i.e., water-pumping windmills, aero-turbines (a small wind electric generator having a ability of as much as 30 kW) and wind–sun hybrid systems [16].

3.4. Solar Energy

Sun electricity is the most ample permanent power resource on the earth and it is available to be used in its direct (sun radiation) and indirect (wind, biomass, hydro, ocean, etc.) forms. Solar strength, experienced by using us as warmness and light, may be used via routes: the thermal path makes use of the heat for water heating, cooking, drying, water purification, energy generation, and other programs; the photovoltaic route converts the light in solar energy into strength, that may then be used for some of functions inclusive of lighting, pumping, communications, and energy deliver in un electrified regions. The whole annual sun radiation falling on the earth is more than 7500 times the arena's overall annual primary power intake of 450 EJ. The once a year solar radiation reaching the earth's surface, about three,400,000 EJ, is an order of magnitude greater than all the expected (discovered and undiscovered) non-renewable strength resources, which include fossil fuels and nuclear. However, 80% of the existing international strength use is based on fossil fuels. Maximum components of India receive 4–7 kWh of solar radiation according to rectangular meter in step with day with 250–300 sunny days in 12 months. The best annual radiation electricity is obtained in Western Rajasthan even as the North- Japanese location of the usa gets the lowest annual radiation.

India has a great level of solar radiation, receiving the solar power equivalent of extra than 5000 trillion kWh/year. Relying at the region, the daily prevalence degrees is around from 4-7 kWh/m2, with the hours of sunshine ranging from 2300 to 3200 per 12 months. The MNRE, running along with the Indian Renewable energy improvement enterprise (IREDA) to promote the utilization of all sorts of solar energy in addition to to increase the percentage of renewable power within the Indian market. This promoting is being done thru R&D, demonstration projects, government subsidy programs, and additionally personal region projects. The prime Minister launched the countrywide action Plan on weather alternate (NAPCC) on 30th June, 2008. The Plan proposes to start eight missions, amongst which one is the countrywide solar assignment [16]. Solar thermal and solar photovoltaic technology are both encompassed by the solar strength program that is being Implemented by way of the Ministry (appeared as certainly one of the most important inside the world) to utilize India's estimated sun strength ability of 20 and 35 MW/km2 solar thermal. India's general potential for solar water heating systems has been predicted to be 140 million m² of collector area. A government scheme for 'increased improvement and deployment of solar Water Heating systems in home, commercial and industrial sectors' has been brought, with the object of selling the set up of any other million m² of collector vicinity at some point of FY 2005–2006 and 2006–2007. The scheme offers a number of economic and promotional incentives; alongside Other measures of aid. Sun air heating technology has been implemented to various business and agricultural procedures (e.g. Drying/curing, regeneration of dehumidifying sellers, wood seasoning, and leather tanning) and additionally for space heating; many styles of sun dryers were advanced for use in exceptional conditions. The authorities present financial support for solar air Heating/drying structures, and additionally for solar concentrating systems. Sun buildings had been promoted by the MNRE that allows you to boom strength performance; the nation government in Himachal Pradesh has actively promoted the incorporation of passive sun layout into building layout. The solar Photovoltaic program (SPV) promoted with the aid of the Ministry for the past decades, has been aimed particularly at rural and far off areas. Following the fulfillment of the United states-huge SPV demonstration and usage program for the duration of the length of the 9th Plan, it's miles deliberate, with sure changes, to retain it throughout the tenth Plan (2002–2007).The Ministry has the goal that with the aid of 2010 they may all have get right of entry to to strength from renewable strength assets [9].

3.5. Geothermal Energy

Geothermal is electricity generated from warmth stored inside the earth, or the collection of absorbed warmness derived from underground. Titanic quantities of thermal energy are generated and saved in The Earth's center, mantle and crust. Geothermal strength is at gift contributing approximately 10,000 MW over the world and India's small sources can augment the above percentage. Research Carried out with the aid of the geological survey of India have discovered life of about 340 hot springs within the hot US. Those are distributed in 7- geothermal provinces. The provinces, although determined along the west coast in Gujarat and Rajasthan and along a west-south west-east-northeast line jogging from the west coast to the western border of Bangladesh (called SONATA), are most prolific in a 1500 km stretch of the Himalayas. The aid is little used at the moment but the authorities has an formidable plan to extra than double the cutting-edge total set up Generating ability via 2012.

4. OTHER RENEWABLE ENERGY TECHNOLOGIES THE CLEAN FACTS

Sun thermal technologies, especially solar water heating system, sun cookers and solar generation systems are the maximum commercialized technologies amongst renewable strength technology in India. Regulations are set to provide further impetus to dissemination of sun technology. Biogas represents an alternative supply of energy, derived specifically from organic wastes. In India, the use of biogas derived from animal waste, frequently cow dung has been promoted for over 3 a long time now. Biogas is a smooth fuel produced thru anaerobic digestion of a spread of natural wastes: animal, agricultural, home, and business. Biogas is the best technology that has placed cooking in rural areas on technological ladder and has made cooking a satisfaction with associated social and environmental blessings including 0 indoor pollutants. India's country wide challenge on Biogas development (NPBD) has been one of the nicely prepared and systematic applications to provide logistic and institutional support for that has been underneath implementation since early 1980s. India Biogas software is one of the maximum successful software if we evaluate with different such application implemented in Rural India. Until December 2004, below the countrywide Biogas software, over 3.7 million biogas plants inside the capacity of one–6 m3 had been installed. The remaining aim of this program is to set up biogas flora in around 12 million families that have sufficient livestock to maintain a everyday supply of dung. Biofuels program inside the US is at nascent stage. The policy measures presently in region encompass an excise tax discount for E-5,The responsibility to combo all petrol with 5% ethanol in positive regions due to the fact January 2003 and authorities regulation of the ethanol promoting charge on the basis of ethanol manufacturing fees. Eventually the percentage of ethanol combination in petrol is planned to be increase to 10%. New biofuels coverage for the US is under construction. Hydrogen strength is likewise at early degree of development. Ministry of new and Renewable strength also funded research projects on exceptional factors of hydrogen energy technology development. India is the member of the global Partnership on Hydrogen financial system (IPHE) set up in Washington, DC in November 2003.Destiny demanding situations to India consists of reducing value of hydrogen drastically and improve production rates from extraordinary strategies, improvement of compact and cheaper storage Capacity, establishment of hydrogen network and development of hydrogen fuelled IC engine and performance improvement of different type of fuel cellular structures. The street map envisages taking Up of studies, improvement and demonstration sports in various sectors of hydrogen electricity technologies and visualized dreams of one million hydrogen-fuelled automobiles and 1000MW Combination hydrogen based totally energy era potential to be set up in the usa by means of 2020 [17].

5. RENEWABLE ENERGY AND SUSTAINABLE DEVELOPMENT THE CLEAN FACTS

To preserve financial boom and lift living requirements, strength shortages can be met by increasing components. However there are different critical issues: environmental sustainability and social development. The current pattern of economic growth has triggered critical environmental harm – polluting the air, developing massive quantities of waste, degrading organic structures and accelerating weather exchange – with many of those effects coming from the energy quarter. On the same time, it's also critical to don't forget the impact on social development. The dearth of get admission to to power services aggravates many social concerns, together with poverty, ill- health, unemployment and inequity. In modern financial sectors one of the important assets of strength Is oil. Although the arena's biggest oil consumer continues to be America, 4-Asian international locations are not a ways in the back of; China comes second, Japan 1/3, India fourth and the Republic of Korea 6th [18]. Herbal fuel is also increasingly vital: its gasoline performance makes it an attractive desire for new electricity generating plant life and for the economic zone. Different environmental issues include water pollutants and the disposal of waste, particularly nuclear waste. Within the rural areas one worry is the overexploitation of environmentally sensitive regions. Many humans in rural areas rely on biomass fuels for cooking, heating and lighting fixtures. Overuse of these can result in degradation of watersheds, and loss of biodiversity and habitats. Approximately 70% of total greenhouse gas (GHG) emissions are related to energy, in particular from the combustion of fossil fuels for warmth, power technology and shipping. Countries have many alternatives for lowering GHG emissions-at minimum, zero or even internet poor charges. These include power conservation along side will increase in performance, higher energy management, cleaner manufacturing and consumption, and modifications in existence. Renewable and other more green technologies would also assist mitigate weather alternate. Standard, international locations can foster technology-primarily based selection-making that creates incentives for cleaner and more energy-green financial activities while increasing humans get admission to modern power offerings. Renewable electricity has an immediate link to sustainable improvement via its effect on human improvement and financial productivity [19]. Renewable electricity resources offer possibilities in terms of power safety, social and economic improvement, get right of entry to energy, mitigation of weather trade and decreasing the impact at the environment and health. In Figure 6 we will see the opportunities of renewable electricity sources toward sustainable improvement.

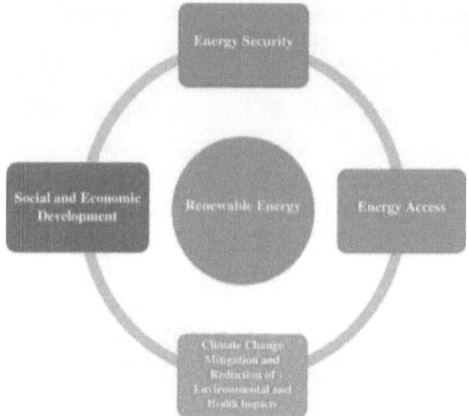

Figure 6: Renewable energy the Clean Facts

5.1. Energy Access

The Sustainable Improvement Goal-7 (clean and cheap electricity) pursuits to make certain that energy is smooth, on hand, available and on hand to all, and this could be finished with renewable strength as they may be normally disbursed around the globe. Get right of entry to to worries have to be understood in a nearby context, and in most international locations there may be an obvious distinction between electrification in city and rural regions, in particular in Sub-Saharan Africa and in the South Asian region [20]. Renewable grid-primarily based grids are typically extra aggressive in rural areas with big distances to the national grid and coffee degrees of rural electrification, are widespread openings for renewable strength-primarily based mines to ensure their get right of entry to power.

5.2. Energy Security

The notion of power security is typically used, but there's no consensus on its specific interpretation. Nevertheless, the concern inside the discipline of strength protection is based on the concept that there is a non-stop deliver of power this is essential to the functioning of an economy [21]. Thinking about the interdependence among monetary boom and power consumption, get entry to to a solid supply of energy is essential to the political international and a technical and economic challenge for both developed and growing nations, as prolonged interference would cause extreme financial and basic functioning problems for most societies [22]. Renewable energy assets are disbursed calmly across the globe as compared to fossils and, in well known, less traded in the marketplace. Renewable power reduces strength imports and contributes to diversifying the delivery options portfolio and lowering the vulnerability of the economic system to charge volatility and represents opportunities to strengthen electricity protection across the globe. The introduction of renewable energy can also help to increase the reliability of specific power services in areas that often suffer from inadequate get entry to to the grid. A diverse portfolio of energy resources alongside good control and system design can help to increase safety [23].

5.3. Social and Economic Development

Typically, the strength quarter has been perceived as a key to financial improvement with a sturdy correlation among financial boom and expansion of electricity consumption. Globally, in keeping with captain comes are undoubtedly correlated with per capita energy use and financial increase can be recognized as the most essential component at the back of growing power intake in the ultimate many years. It in flip creates employment; renewable electricity examine in 2008, proved that employment from renewable energy technology become about 2.3 million jobs global, which also has stepped forward fitness, training, gender equality and environmental safety [23].

5.4. Climate Change Mitigation and Reduction of Environmental and Health impacts

Renewable electricity assets used in energy manufacturing make a contribution to reducing greenhouse gas emissions that mitigate climate trade, reduce environmental and fitness headaches related to pollutants from fossil strength sources that the exchange in total GHG emissions within the eco-surroundings. The EEA (1990-1990) and GHG in step with capita emissions are shown in Figures 2 and 3. Discern 2 shows that greenhouse gas emissions decreased by means of 14% between the 33 EEA countries 1990-2012. However, there were variations in individual nations, even as there are Right here changed into a decrease in GHG emissions in 22 EEA nations, there has been an increase in eleven EEA nations. GHG emissions per capita reduced by means of 22% between 1990-2012 in the EEA countries, as described in parent three (EEA, 2016).

5.5. Climatic Changes

Climatic changes, because of global warming as a result of greenhouse gases, especially carbon dioxide (CO_2) produced throughout the burning of fossil fuels, have been causing massive modifications inside the ecosystems and main to nearly a 150,000 extra deaths each year [3]. This upward push is especially caused by the unsustainable use of fossil fuels and the adjustments in the use of the land [24].

5.6. Clean Development Mechanism

The clean development mechanism (CDM) of the Kyoto Protocol has been installation to assist developing international locations in achieving sustainable development via promoting greenhouse gasoline emission reduction initiatives, that generate emission credit (certified emissions discounts, CERs) for industrialized countries [25]. A number of nations inside the place are taking gain of the CDM. This is a provision of the Kyoto Protocol which was devised firstly as a bilateral mechanism via which entities in industrialized international locations ought to gain licensed emission discounts (CERs) by using investing in easy technologies in growing nations. For the recipient growing countries, this will boost returns on initiatives by up to 12% for wind, hydro and geothermal tasks and by using 15–17% for biomass and municipal waste tasks (UNEP).Indian organizations have already dedicated funding to generate greater than 379 million CERs. Worldwide investments have been made so that it will generate 1.9 billion CERs by means of 2012.

6. FUTURE OF RENEWABLE ENERGY CLEAN FACTS IN INDIA

India, confronted with twin challenges on power and environmental the front, has no option however to work in the direction of increasing the position of renewable in the destiny electricity structures. Renewable electricity Technologies range broadly in their technological adulthood and business fame. In India, renewable electricity is on the take-off stage and corporations, industry, government and customers have a huge variety of problems to cope with before those technologies should make a real penetration. India with huge renewable electricity sources (sun PV, wind, solar heating, small hydro and biomass) is to set to have huge-scale improvement and deployment of renewable strength initiatives [26]. The aim of assembly 10% of the US Electricity deliver through renewable by 2012 and additionally bold plans for the distribution of biogas plants, solar PV packages and Sun City seems to be within attained. Furthermore, introduction of tradable renewable power certificates (REC) could Triumph over the existing gap that is hindering the application of quota for renewable and thereby creates a colorful market place. India would additionally have to search for worldwide cooperation in renewable strength through properly defined R&D projects with proper division of labor and responsibilities for specific obligations with equitable financial burden and credit sharing arrangements. Renewable power development is considered in India to be of fantastic importance from the factor of view of lengthy term energy supply safety, environmental benefits and climate alternate mitigation. The included energy coverage record has recognized the need to maximally increase home supply options as well as the want to diversify strength resources. The Committee has placed emphasis on Higher use of renewable in all types of offerings. It's far anticipated that the contribution from renewable in electricity era alone may be of the quantity of 60,000MW inside the year 2031–2032. Via 2031–2032 renewable will be the key motive force in social inclusion of the negative inside the improvement process. A modest evaluation of investments in the renewable power quarter may be about Rs. 300,000 crores over the subsequent 25 years. MNRE has blanketed in its undertaking: electricity security; growth in the percentage of clean power; power availability and get admission to; Electricity affordability; and strength fairness [27]. Some of authorities and personal organizations which include MNRE, Centre for Wind energy generation, Universities, IITs, NITs, Indian Oil enterprise Ltd. (IOCL) and The electricity resource Institute (TERI) are worried in R&D of renewable strength resources.

7. CONCLUSIONS

Electricity is a compulsory asset in our everyday life as a manner to improve human improvement that results in increase and productiveness. Returning to renewable assets will help mitigate climate change but must be sustainable to ensure a Sustainable Future for generations to satisfy their energy wishes. Understanding of the connection among sustainable development and, especially, renewable power continues to be restrained. The aim of the paper changed into to decide whether it's miles feasible to regenerate strength resources which can be sustainable and the way switching from fossil strength resources to renewable sources of electricity might make contributions to reducing climate trade and its effect. Qualitative studies have been used to review the articles within the scope of the take a look at.

However, the overall existence cycle of renewable electricity assets does no longer have internet emissions to assist restriction the future worldwide greenhouse gas emissions. But, fee, price, political environment and market conditions have end up the barriers that save you growing nations, the least advanced and the developed ones from making full use of their ability. In this way, developing a international possibility via global cooperation to help LDCs and developing nations in terms of accessibility of renewable energies, energy performance, smooth green technologies and studies and investment in energy infrastructure will lessen the value of renewable electricity, put off boundaries to power efficiency (high upgrading charge) and sell new ability for mitigating weather alternate. The study highlighted opportunities for renewable energy resources; power safety, access to energy, social and economic improvement and mitigation and discount of weather alternate affects on the environment and on fitness. There are challenges that tend to bog down the sustainability of renewable strength resources and its potential to mitigate climate alternate. These challenges are: market disasters, lack of data, get admission to to raw materials for the destiny use of renewable sources, and, most significantly, our way of the use of strength inefficiently. Based totally on the findings, the subsequent suggestions may be made that may assist improve renewable electricity issues as being sustainable and also reduce the ozone depletion charge because of greenhouse gasoline emissions, especially carbon dioxide (CO_2):

Electricity safety, financial boom and environment protection are the national strength policy drivers of any United States of America of the sector. The need to reinforce the efforts for in addition improvement and merchandising of renewable power sources has been felt world over in light of excessive fees of crude oil. A crucial part of the solution will lie in selling renewable energy technology as a way to deal with worries approximately electricity protection, monetary boom within the face of rising strength prices, competitiveness, fitness expenses and environmental degradation. In step with NAPCC other resources of renewable power would be promoted. Specific movement points that have been stated include selling deployment, innovation and basic research in renewable energy technology, resolving the obstacles to development and business deployment of biomass,

Hydropower, sun and wind technologies, selling instantly (direct) biomass combustion and biomass gasification technologies, selling the improvement and manufacture of small wind Electric powered mills, and improving the regulatory/tariff regime if you want to most important movement renewable electricity resources inside the country wide strength gadget. For that reason, expanded awareness is being laid on the deployment of renewable strength this is likely to account for around 5% within the strength-blend by means of 2032. Alternate fuels, basically bio-fuels, are proposed to be steadily used for mixing with diesel and petrol, in particular for shipping applications. In the end, renewable energy gives great advantages and can make contributions notably in the countrywide power blend as a minimum financial, environmental and social charges and it's far expected that the proportion of renewable energy in the general technology ability will growth in destiny.

REFERENCES

[1] Varuna SK, Singal. Review of augmentation of energy needs using renewable energy sources in India. Renewable and Sustainable Energy Reviews

[2] International Energy Agency IEA. Key world energy statistics. Available at: http://www.iea.org/Textbase/nppdf/free/2006/Key2006.pdf [Accessed: 07/06/2007].

[3]   World Energy Outlook. International energy agency; 2008. http://www.worldenergyoutlook.org/ 2008.asp.

[4]   REN21, Renewables 2007 global status report. http://www.ren21.net/pdf/.

[5]   REN21, Renewables 2009 global status report. http://www.ren21.com.

[6]   A. Verbruggen, „Renewable energy costs, potentials, barriers: Conceptual issues," Energy Policy, pp. 850-861, 2010. 2007;11:1607–15.

[7]   Planning Commission, Govt. of India—September 1995 & September 1996 Projections to 2020–2021.

[8]   Subramanian V. Renewable energy in India: status and future prospects.Ministry of New and Renewable Energy; November 2007.

[9]   GOI. Tenth Five year plan 2002–2007, planning commission, New Delhi. Available at: http:// planningcommission.nic.in/aboutus/committee/wrkgrp11/wg11_ renewable.pdf.

[10]   Urja Akshay. Newsletter of the Ministry of New and Renewable Energy,Government of India; October 2008. http://mnes.nic.in/akshayurja/sept-oct-2008-e.pdf.

[11]   India 2009. Energy Publication Division. Ministry of Information & Broadcasting Government of India; 2009.

[12]   Senneca O. Kinetics of pyrolysis, combustion and gasification of three biomass fuels. Fuel Process Technology 2006;87–97.

[13]   Ramachandra TV, Kamakshi G, Shruthi BV. Bioresource status in Karnataka. Renewable and Sustainable Energy Reviews 2004;8:1–47.

[14]   Bridgwater AV, Toft AJ, Brammer JG. A techno-economic comparison of power production by biomass fast pyrolysis with gasification and combustion. Renewable and Sustainable Energy Reviews 2002;6:181–246.

[15]   KPMG. India energy outlook; 2007.

[16]   Urja Akshay. Newsletter of the Ministry of New and Renewable Energy. Government of India; December 2008. http://mnes.nic.in/akshayurja/novdec-2008-e.pdf.

[17]   Ghosh D, Shukla PR, Garg A, Ramana VP. Renewable energy technologies for the Indian power sector: mitigation potential and operational strategies. Renewable and Sustainable Energy Reviews 2002;6:481–512.

[18]   Conn I. Energy trends and technologies for the coming decades. Address to the Harvard University Center for the Environment; 2007.

[19]   Asumadu-Sarkodie, Feasibility of biomass heating system in Middle East Techinical University, Northern Cyprus campus: Cogent Engineering, 2016.

[20]   A. Brew-Hammond, „Energy access in Africa: Challenges ahead," Energy Policy, pp. 2291-2301, 2010

[21]   Kruyt, „Indicators for energy security," Energy, nr. 37, pp. 2166-2181, 2009.

[22]   N.K. H.H. Larsen, „How do we convert the transport sector to renewable energy and improve the sector's interplay with the energy system?," în Transport renewable energy in the transport sector and planning, Copenhagen, Technical University of Denmark, 2009, pp. 61-94.

[23] Edenhofer, Reneable Energy.

[24] Intergovernmental Panel on Climate Change—IPCC. Cambio clima'ticoy biodiversidad' Working Group II report; 2001. Available in: http://www.ipcc.uch. Accessed: 10/05/07.

[25] Purohit P, Michaelowa A. CDM potential of SPV pumps in India. Renewable and Sustainable Energy Reviews 2008;12:181–99.

[26] Maithani PC. Renewable energy policy framework of India. India: Narosa Publication Delhi; 2008. p. 41–54.

[27] Chaturvedi P, Garg HP. Financing renewables—emerging dimensions. IREDA NEWS; July–September 2007. http://www.ireda.in/pdf/July-September_2007.pdf.

[28] A. Kumar et al. / Renewable and Sustainable Energy Reviews 14 (2010) 2434–2442

10

Theoretical Investigation of Geometry Optimization Vibration Analysis Thermodynamical Properties Electronic Properties of Hexachlorophosphazene: A DFT study

Vijay Narayan Mishra[a], D.V Shukla[b], Ankit Kumar Sharma[c], Ashutosh Kumar Singh[d], Vinod Kumar Singh[e], Shubham Tiwari[c], Anoop Kumar Pandey[*c]

[a]S.R.M.G.P.C. Lucknow (India)
[b]GLA university Mathura U.P.
[c]K. S. Saket Post Graduate College, Department of Physics
[d]K. S. Saket Post Graduate College, Department of Chemistry
[e]Member of UPHESC Prayagraj
*Email: anooppandeyias@gmail.com

ABSTRACT

In present communication geometry optimization of Hexachlorophosphazene has been carried out by combination of DFT/B3LYP method and 6-311G (d,p) basis set. The vibrational analysis of Hexachlorophosphazene is calculated on its optimized geometry. The assignment of vibrational modes frequencies and its intensities are calculated by using same level theory. The electronic properties of title molecule has been discussed by using several chemical reactivity parameters. The nature and reactivity sites are calculated by using HOMO LUMO MESP plot. The electronic transition in title molecule has been calculated by time dependent theory (TDDFT) on same level theory. The thermodynamical parameters are calculated on variation of temperature (100K-600K).

Kay words: DFT, HOMO, LUMO, TDDFT

1. INTRODUCTION

The chemical formula with $(NPCl_2)_3$ molecule is known as Hexachlorophosphazene. The cyclic, unsaturated support containing of irregular phosphorus and nitrogen centresmolecule has can observed as a trimer of the theoretical compound $Na''PCl_2$. Its organization as a phosphazene underlines its connection to benzene[1]. A large academic interest incorporate with phosphorus-nitrogen bonding and phosphorus reactivity[2-3].

Infrequently, profitable or recommended applied submissions have been described, too, utilising hexachlorophosphazene as a forerunner chemical[Offshoots of well-known attention contain thehexalkoxyphosphazene emollientsfound from nucleophilic substitution of hexachlorophosphazene with alkoxides[4], or chemically unaffected inorganic polymers with necessarythermal as well asmechanical properties known as polyphosphazenes twisted from the polymerisation of hexachlorophosphazene[2,4].

The hexachlorophosphazene was firstly synthesis by von Liebig in 1834. The experiments of synthesis conducted by Wöhler[5]. The reaction has made in between phosphorus pentachloride and ammoniaexothermically and a new compound obtainedin which toto remove the ammonium chloride be eroded with cold water coproduct. The newly compound obtained which having P, N, and Cl, atoms. It was subtleto hydrolysis by hot water [2].

Novel synthesis which depends onprogresses by Schenk and Römer who utilized ammonium chloride rather than ammonia and alsopassive chlorinated solvents. The replacement of ammonium chloride in place of ammonia permits the feedback to continuehaving no strong exotherm with NH_3/ PCl_5 reaction. The stand the hydrogen chloride side creation which individual chlorocarbon solvents are 1,1,2,2-tetrachloroethane or chlorobenzene.Meanwhile ammonium chloride is inexplicable in chlorinated solvents, diagnosis is simplified [5-6].

Now a day fast growing computation technique quantum chemical calculation are useful tool to calculate geometry optimization information about transition compound FTIR analysis, Optical spectra UV-Vis, nonlinear optical properties(NLO) electronic properties, Thermodynamical properties, Natural bond analysis (NBO) of any chemical system. Density functional theory (DFT) along with exchange functional(B3LYP) is useful method to calculate various static as dynamical properties of any chemical system.

In ongoing communication [7-11] geometry optimization vibrational analysis, electronic properties along with optical spectra UV-Vis by using TDDFT of Hexachlorophosphazene are calculated by using DFT/B3LYP method. The thermodynamical properties of Hexachlorophosphazene are calculated by using variation of temperature(100K-600K). The calculated correlation equation provides new path to synthesize Hexachlorophosphazene. In best of knowledge no such type of study conducted elsewhere till now.

2. COMPUTATIONALDETAILS

The title molecule is designed by using Gauss view and after that geometry optimized by using combination of DFT [12-13] and B3LYP[14-15] method without any symmetric constrains. The whole calculation has been done by using Gaussian 16 program package [16] on our personal laptop. The vibrational analysis is calculated by using gauss view 6.0 program package [17]. The electronic properties are calculated by using HOMO LUMO MESP plots and various quantum chemical descriptor. The UV-Vis optical spectra of title molecule is calculated by using time dependent theory TDDFT. The UV-Vis plot of title molecule is drawn from gauss sum 3.0.

3. RESULTS AND DISCUSSION

3.1 Optimized Structure

The optimized geometry of title molecule having no symmetry but this shows C1 symmetry. The ground state energy of title molecule having -2211.54a.u. The animated gassview shows that title molecule replaced three C atom with three Nitrogen and other three carbon replaced by phosphorus P atoms in alternate position. The two Cl atoms attach with P9 atom having bond length in between $2.038A^0$-$2.032A^0$,The Cl atom attach with P7 with bond length in between $2.038A^0$-$2.037A^0$ and attached P8 atom two bond length lies in between $2.037A^0$-$2.038A^0$. The P-N bond length in ring lies in between $1.405A^0$-$1.408A^0$. The bond angle in between <PNP lies in between 116.90^0-123.10^0. The bond angle <Cl-P-Cl lies in between 67.22^0-71.42^0. The optimized structure is shown in Figure 1.

3.2 Vibrational Analysis

The title molecule having 12 atoms(N) so having (3N-5) 31modes of vibrations. In these modes 30 (N-1) modes are stretching modes and rest are bending modes. The whole spectra are ranging for

117-1136 Cm^{-1} which are basically lies in finger print region. The whole spectra lie to lower frequency region due their heavy mass. The calculated frequencies are overestimated due to ignorance of electron-electron correlation, anharmonicity, molecule-molecule interaction so to compare calculated frequencies with observed frequencies calculated frequencies are scaled with 0.96. The assignment and scaled wavenumbers along with their intensities are listed in table-1.

Below, we discuss and analyse the significant modes of vibration in the molecules under study. Scissoring and twisting vibrationmodes are caused by the attached PCL2 group to the ring in the titlemolecule.Other vibrations, such as in-plane and out-of-plane bendingcan be seen at lower frequencies. In vibrational analysis frequency range varies 117.94 cm^{-1}to 1136.36 cm^{-1}.At higher frequencies region an intense mode appears at 1136.36cm^{-1}due to R[τ_i(CL8-P1-N5-P3)]. The other intense polarized modes appear at 1071.89cm^{-1} due to R[τ_i(N6-P3-N5-CL11)]. The intense mode appears at994.94cm^{-1} due to R[τ_i(CL12-P3-N5-CL11)]. In pane scissoring modes of vibration appears at 578.51cm^{-1}R[Γ(CL7-P1-N4)].The intense symmetric intense stretching modes at 365.79cm^{-1}, 326.28cm^{-1} due toR[$<_s$(CL10-P2)] andR[$<_s$(CL9-P2)] respectively.

4. HOMO-LUMO AND MESP SURFACES

The frontier molecular orbitals (HOMO-LUMO) and their properties such as energy are very useful for physicist and chemists and are very important parameters for quantum chemistry. This is also useful for predicting the most reactive position in pie-electron system and also explains several types of reaction in conjugated system [18-19]Now these energy gaps are used to prove the bioactivity from intra-molecule charge transfer. The HOMO and LUMO plot of molecule is shown in Figure. 2.

The importance of molecular orbital potential (MESP) lies in the fact that it simultaneously displays molecular size and shape as well as positive, negative and neutral electrostatic potential regions in form of colour grading, which is very helpful in the investigation of the most probable binding receptor site along with the size and shape of the molecule [20-23]. The MESP plot of molecule is also displayed in Figure. 3.

4.1 Electronic parameters

To describe the electronic properties of molecule, we have calculated various electronic parameters like ionization potentials (I), electron affinity (A), absolute electronegativity (χ) and chemical hardness (η) etc. X and η can be calculated by using finite –difference approximations [24-27] as η = ½ (I-A) and χ = ½ (I+A). These parameters often used to describe chemical reactivity of molecules are listed in Table 2.

4.2 UV-Vis analysis

The optical transition in title molecule is calculated by using time dependent theory (TDDFT) by using same level theory on optimized geometry. The calculated UV-vis sepectra of title molecule is shown in fig- The Calculated optical transition electronic transitions: E (eV), oscillatory strength (f), λ_{max} (nm) transition orbital and their % contributions are using TD–DFT/B3LYP/6–311G(d,p) method. In UV-vis spectra one prominent peak appears at 230nm with oscillatory strength (f=0.0103) having transition energy is corresponding to transition between $S_0 \rightarrow S_3$. The orbital contribution in this transition are H-7$\rightarrow$LUMO (36%), H-6$\rightarrow$L+1 (15%), H-5$\rightarrow$L+2 (15%), H-2'!LUMO (15%). The other two peak with same sharpness appears at 216nm and 217nm with (f=0.0049) are having transition energy 5.724, 5.723eV are transition corresponding $S_0 \rightarrow S_7$, $S_0 \rightarrow S_8$ respectively. The molecular orbital contribution in between these transitions are H-4$\rightarrow$LUMO (35%), H-3$\rightarrow$LUMO (30%) and H-4$\rightarrow$ LUMO (30%), H-3$\rightarrow$LUMO (35%) respectively.

4.3 Thermodynamic Parameters

The various calculated thermodynamical parameters like entropy, enthalpy specific heat by using vibrational analysis. The vibrational energy contributes more than 90% thermal energy. We have calculated these parameters with variation of temperature (100-600K). Theentropy, enthalpy specific heat are increase with increases temperatures (table-4). The calculated correlation equation along with polynomial of order two are

$$H^0_m = 16.1561 + 0.03981\ T^1_m + 0.00009411 T^2_m \qquad \text{for}(R^2 = 0.98521$$

$$Cv^0_m = 13.0796 + 0.13287 T^1_m - 0.00009435 T^2_m \qquad \text{for}(R^2 = 0.97312)$$

$$S^0_m = 58.1716 + 0.28651 T^1_m - 0.000185243\ T^2_m \qquad \text{for}(R^2 = 0.9865)$$

The calculated correlation factor for entropy $R^2 = 0.98521$, enthalpy ($R^2 = 0.97312$) specific heat for ($R^2 = 0.9865$) shows a good correlation in between temperature and these thermodynamical parameters.The calculated correlation equation gave new chemical path to synthesize Hexachlorophosphazene.

5. CONCLUSION

This study presents a complete computational structural study on molecule. All calculated wave numbers are real in nature for all the molecules thus all the compounds are stable.All the vibrational assignments have been done for the first time by employing combination of DFT/B3LYP and 6-311G(d,p) basis set. Detailed descriptions of the vibrational spectra of these molecules have been done with the help of Normal modes analysis. Reactivity plays a key role in specific chemical reactions. We have also discussed sites for both molecules during electrophilic, nucleophilic, and radical attacks with the help of global reactivity descriptors. The findings from this research can provide a suitable path for researchers in future.

6. ACKNOWLEDGMENTS

One of the authors, Anoop Kumar Pandey, is grateful and thanks to the Uttar Pradesh government (India)[No:46/2021/603/sattar-4-2021-4(56)/2020] for providing him financial support.

REFERENCES

[1]   C. X. H. Choi, M. Kertz, J. Phys. Chem. A 101 (1997) 3823.

[2]   C. Ravikumar, I. H. Joe, V.S. Jayakumar, Chem. Phys. Lett. 460 (2008) 552.

[3]   Pirnau, V. Chis, O.Oniga, N, Leapold, L. Szabo, M. Baias, O. Cozar, Vib. Spectrosco. 48 (2008) 289.

[4]   J. S. Murray, K. Sen, Molecular Electrostatic potential; Concepts and Applications, Elsevier Amsterdam, Netherland (1996).

[5]   J. Sponer, P. Hobza, Int. J. Quantum Chem. 57 (1996) 959.

[6]   R.G. Parr, W.Yang, Density functional theory of atoms and molecules, Oxford Univ. Press and Clarendon press, New York and Oxford (1989).

[7]   A K Pandey, S A Siddiqui, ADwivedi, NMisra, Kanwal Raj, FTIR spectra and Vibrational Spectroscopy of Loganin using Density Functional Theory, Spectroscopy 25 (2011) 287–302.

[8]   S A. Siddiqui A K Pandey,T. Rasheed, M. Mishra Quantum chemical study of RhFnnano clusters: An investigation of superhalogen a in Journal of Fluorine Chemistry Volume 135, March 2012, Pages 285–291.

[9]   N.Misra, S.A Siddiqui, R. Srivastava,A.K.Pandey, S. Jain, Vibrational analysis of boldine hydrochloride using QM/MM approach,.*Spectroscopy Vol. 24 No. 5, (2010) 483-499.*

[10]   N. Misra, O. Prasad, L. Sinha, A. K.Pandey Molecular structure and vibrational spectra of 2 formyl benzonitrile by dendity functional theory and ab-initio Hartree-Fockcalculations, Journal*of Molecular Structure: THEOCHEM 822 (2007) 45-47.*

[11]   S.A. Siddiqui, A.K. Pandey, P. K. Singh, T.Hasan, S. Jain, N. Misra Molecular structure, vibrational spectra and potential energy distribution of colchicine using ab initio and density functional theory", *Journal of Computer Chemistry, Japan Vol. 8, No. 2 (2009) 59-72.*

[12]   P.Hohenberg P, Kohn W, Inhomogeneous electron gas. Phys Rev 136(1964)B864–B871. https://doi.org/10.1103/PhysRev.136.B864

[13]   A.D. Becke,Density functional thermochemistry-III The role of exact exchange, J. Chem. Phys. 98 (1993) 5648–5652. https://doi.org/10.1063/1.464913

[14]   C.T. Lee, W.T. Yang, and R.G.B. Parr, Development of the Colle-Salvetti Correlation-Energy Formula into a Functional of the Electron Density Phys. Rev. 37 (1988) 785–789. http://dx.doi.org/10.1103/PhysRevB.37.785

[15]   G.A. Petersson, and M.A. Allaham, A complete basis set model chemistry. II. Open shell systems and the total energies of the first row atoms, J. Chem. Phys., 94 (1991) 6081–6090.https://doi.org/10.1063/1.460447

[16]   Frisch MJ, Trucks GW, Schlegel HB, Scuseria GE, Robb MA, Cheeseman JR, Scalmani G, Barone V, Petersson GA, Nakatsuji H, Li X, Caricato M, Marenich AV, Bloino J, JaneskoBG, Gomperts R, Mennucci B, Hratchian HP, Ortiz JV, Izmaylov AF, Sonnenberg JL, Williams-Young D, Ding F, Lipparini F, Egidi F, Goings J, Peng B, Petrone A, Henderson T, Ranasinghe D, Zakrzewski VG, Gao J, Rega N, Zheng G, Liang W, Hada M, Ehara M, Toyota K, Fukuda R, Hasegawa J, Ishida M, Nakajima T, Honda Y, Kitao O, Nakai H, Vreven T, ThrossellK,Montgomery Jr JA, Peralta JE, Ogliaro F, Bearpark MJ, Heyd JJ, Brothers EN, Kudin KN, Staroverov VN, Keith TA, Kobayashi R, Normand J, Raghavachari K, Rendell AP, Burant JC, Iyengar SS, Tomasi J, Cossi M, Millam JM, Klene M, Adamo C, Cammi R, Ochterski JW, Martin RL, Morokuma K, Farkas O, Foresman JB, Fox DJ (2016) Gaussian 16, Revision B.01. Gaussian, Inc., Wallingford.

[17]   R. Dennington,T.A. Keith, J.M. Millam (2016) GaussView, Version 6.1. Semichem Inc., Shawnee Missi.

[18]   M. Gutowski, G. Chalasinski,*J. Chem. Phys.*98, 4540–4554(1993

19]   S. C. Bose, H.Saleem, Y.Erdogdu, G. Rajarajan, V.Thanikachalam, *Spectrochim.Acta part A*82, 260–269(2011).

[20]   S.R. Ggadre and R.K. Pathak, Miximal and minimal characteristics of molecular electrostatic potentials, J. Chem. Phys. 93 (1990), pp. 1770–1774.

[21]   S.R. Ggadre and I.H. Shrivastava, Shapes and sizes of molecular anionsviatopographicalanalysisofelectrostaticpotential, J.Chem. Phys. 94 (1991), pp. 4384–4390.

[22]   J.S. Murray and K. Sen, Molecular Electrostatic Potentials, Concepts and Applications, Elsevier, Amsterdam, 1996.

[23]   I. Alkorta and J.J. Perez, Molecular polarization potential maps of the nucleic acid bases, Int. J. Quant. Chem. 57 (1996), pp. 123–135.

[24]   R. G Parr, R.G Pearson, *J. Am. Chem. Soc.* 105, 7512–7516(1983)

[25]   P. Geerlings, F. D.Proft, W. Langenaeker, *Chem. Rev.*103, 1793–1874(2003)

[26]   R.G. Parr, R.A. Donnelly, M. Levy, W.E. Palke, J. Chem. Phys. 68 (1978) 3801.

[27]   R.G. Parr, R.G. Pearson, J. Am. Chem. Soc. 105 (1983) 7512

[28]   R.G Parr, L. Szentpály, S. Liu, *J. Am. Chem. Soc.*121, 1922–1924(1999)

Table-1: Vibrational analysis of prominent modes of vibration frequencies (Cm-1) Intensity title molecule

Calculated Freq(cm^{-1})	Intensity I.R	Assignment
117.94	1.94	R[σ(N6-P2-N4)]
140.85	3.56	R[σ(CL7-P1-N4)]
149.65	0.47	R[τ$_t$(P3-N6-P2-CL10)]
162.26	3.79	R[τ$_t$(P3-N6-P2-CL10)]
175.40	6.78	R[τ$_t$(CL12-P3-N5-CL11)]
190.38	5.42	R[τ$_o$(P3-N5-P1-CL7)]
214.41	5.21	R[τ$_o$(P3-N5-P1-CL7)]
256.86	26.52	R[τ$_o$(P3-N6-P2-CL9)]
273.40	34.66	R[σ(N6-P2-N4)]
284.45	9.82	R[σ(P2-N4-P1)]
323.86	7.89	R[$_s$(CL10-P2)]
365.79	20.85	R[$_s$(CL9-P2)]
389.20	30.39	R[τ$_t$(N6-P3-N5-CL11)]
461.01	13.91	R[τ$_t$(CL11-N5-P1-CL8)]
467.27	43.73	R[τ$_t$(CL8-P1-N5-CL11)]
516.19	80.28	R[σ(CL9-P2-N4)]
578.51	61.78	R[σ(CL7-P1-N4)]
624.30	12.98	R[σ(N6-P2-N4)]
649.73	81.65	R[σ(N4-P2-N6)]
657.82	26.73	R[σ(CL9-P2-CL10)]
687.49	181.64	R[σ(N4-P1-CL7)]
752.17	80.59	R[τ$_t$(CL9-P2-N4-P1)]
957.97	95.09	R[σ(P2-N4-P1)]
994.94	381.49	R[τ$_t$(CL12-P3-N5-CL11)]
1071.89	277.83	R[τ$_t$(N6-P3-N5-CL11)]
1136.36	7.49	R[τ$_t$(CL8-P1-N5-P3)]

Abbreviations: τ$_t$= twisting σ=bending, $_s$=symmetric stretching$_{as}$=antisymmetric stretching=stretching

Table-2: Calculated Electronic Parameter of Molecule

HOMO	LUMO	Gap	χ(eV)	μ(eV)	η(eV)	S(eV)$^{-1}$	ω(eV)
-8.680	-2.857	5.8231	4.3401	-4.3401	2.9115	0.1717	22.859

Table-3: The Calculated optical transition electronic transitions: E (eV), oscillatory strength (f), λ_{max} (nm) using TD–DFT/B3LYP/6–311G(d,p) method

S.N.	Trans. Orb. % cont.	Tans ene. (eV)	Oss. Stre.	λ_{max}(nm)	Assignment	Trans.
1	H-4→LUMO (35%), H-3→LUMO (30%)	5.723	0.0049	217	n_p→R_y*	S_0→S_7
2	H-7→LUMO (36%), H-6→L+1 (15%), H-5→L+2 (15%), H-2→LUMO (15%)	5.382	0.0103	230	n_p→R_y*	S_0→S_3
3	H-4→LUMO(30%), H-3→LUMO (35%)	5.724	0.0049	216	n_p→R_y*	S_0→S_8
4	H-13→LUMO (13%), H-12→LUMO (17%), H-9→L+2(14%), H-3→LUMO (12%)	6.108	0.0672	203	n_p→R_y*	S_0→S_{11}

Table-4: Calculated Thermodynamic parameter of Molecule.

Temperature(k)	E (Thermal) KCal/Mol	CV (Cal/Mol-Kelvin)	S(Cal/Mol-Kelvin)
100	20.393	24.583	84.420
200	23.637	38.974	107.702
300	30.698	40.508	130.698
400	33.029	52.802	141.078
500	38.493	56.244	153.700
600	44.240	58.541	164.534

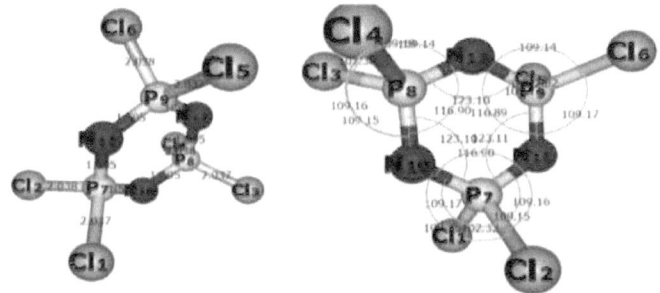

Figure 1: Most stable optimized geometry of title molecule.

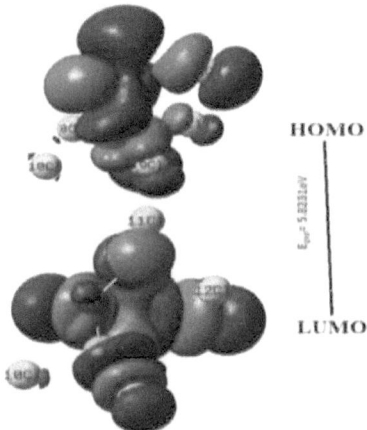

Figure 2: Most stable optimized geometry of title molecule.

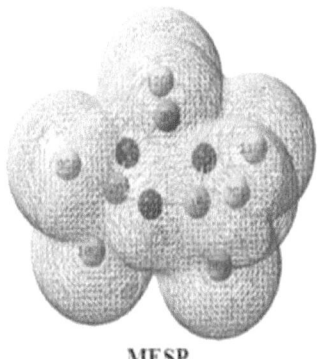

MESP

Figure 3: Molecular electrostatic potential (MESP) surface of title molecule.

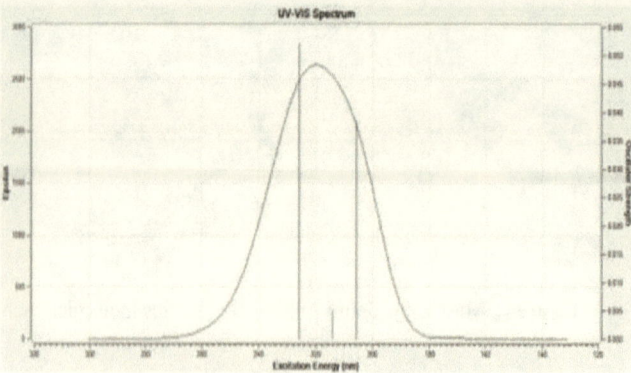

Figure 4: Calculated UV-vis Spectra of title molecule.

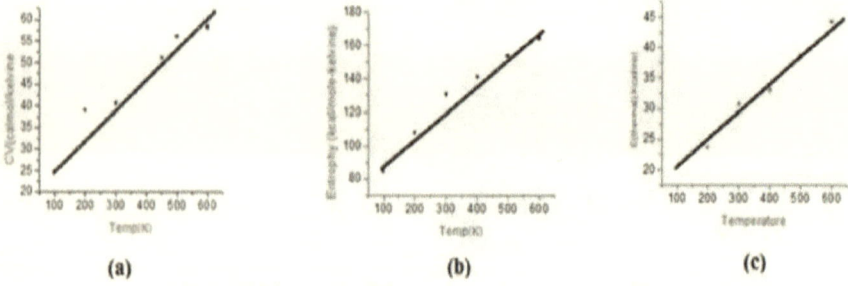

Figure 5: The graph of thermodynamic parameter with temperature.

11

Fuzzy Logic System through Monitoring and Control of Blood Pressure

Amod Kumar Pandey[1], Vikash Kumar Sharma[2*], Abhishek Srivastava[3] and Hemant Kumar[4]

[1,3]Department of Electronics & Communication Engg., School of Management Sciences Lucknow

[2*]Ambalika Institute of Management and Technology, Lucknow

[4]Department of Computer Science & Engg., School of Management Sciences Lucknow

ABSTRACT

In this paper, design and development of a fuzzy logic controller is done for the control of Mean Arterial Pressure (MAP) of a patient during anesthesia. The main purpose of the paper is to constitute a basis for further real time applications. In the simulation studies, the patient is represented with a linear mathematical model that includes time delay elements. The parameters of the fuzzy controller are tuned in order to obtain a robust control performance against the plant parameter variations. Since the modern control is usually realized by using a microprocessor, the discrete time analysis and design of the control system are given in the paper. The simulation results showing the robust control performance of the proposed fuzzy controller under the plant parameter variations are presented.

Keywords: *Fuzzy logic, control, mean arterial pressure (MAP), depth of anesthesia.*

1. INTRODUCTION

The main task of an anesthetist is to control the depth of anesthesia during a surgery in order to provide a painless and comfortable operation for patients [1-3]. However, the depth of anesthesia cannot be easily measured or estimated. The measurement of the Mean Arterial Pressure (MAP), heart rate and pupil size provide some information about the depth of anesthesia. The MAP is used as the most reliable guide for dosing anesthetics by anesthesiologists. The patient's MAP should lie within a predefined range for a good control of depth of anesthesia. The main reason for automating the control of the depth of anesthesia is to release the anesthetist so that he/she can devote his/her attention to other tasks such as controlling the fluid balance, ventilation and drug applications which cannot be adequately automated, thus increasing the patient's safety. Thus, the control of MAP appears as one of the most effective way of controlling depth of anesthesia.

In recent years, fuzzy logic has been used for the control of MAP since the experts (anesthesiologists) knowledge can be Embedded into the controller using fuzzy logic [1, 3].

However, the effects of parameter variations to the control performance of the system are not explicitly considered in most studies. In this paper, the design of a fuzzy logic controller providing a robust control under parameter variations is explained. Simulation results showing the effectiveness of the proposed fuzzy controller under parameter variations are presented.

2. MATHEMATICAL MODEL OF A PATIENT UNDER ANESTHESIA

In this study, the mathematical model of a patient developed in [1] is used for simulation and controller design. The relation between inflow concentration of isoflurane u(t) (input variable) and the resulting MAP y(t) (output variable) is modeled as the sum of two first order terms each with a pure time delay. The model includes the patient and also a semi-closed circuit that is used to deliver the anesthetic agent to the patient. The unit step response is determined a s [1]

$$y(t) = K_1[1 - e^{-a_1(t-\tau_1)}]u(t-\tau_1) + K_2[1 - e^{-a_2(t-\tau_2)}]u(t-\tau_2) \tag{1}$$

Where $K_1 = -3$, $K_2 = -7.3$, $\tau_1 = 23s$, $\tau_2 = 101s$, $a_1 = 0.01$, and $a_2 = 0.006$. The Laplace transform of (1) is

$$Y(s) = K_1[1/s - 1/(s-a_1)]e^{-\tau_1 s} + K_2[1/s - 1/(s-a_2)]e^{-\tau_2 s} \tag{2}$$

And since the input is unit step, i.e., $U(s) = 1/s$, the transfer function between output $Y(s)$ and the input $U(s)$ becomes

$$G_p(s) = \frac{Y(s)}{U(s)} = K_1[1 - s/(s+a_1)]e^{-\tau_1 s} + K_2[1 - s/(s+a_2)]e^{-\tau_2 s} \tag{3}$$

or it can be written as

$$G_p(s) = G_{p1}(s) + G_{p2}(s) \tag{4}$$

Where

$$G_{p1}(S) = \frac{K_1 a_1 e^{-\tau_1 s}}{s + a_1}$$

And

$$G_{p2}(S) = \frac{K_2 a_2 e^{-\tau_2 s}}{s + a_2}$$

In order to obtain a polynomial transfer function and get rid of the time delay elements, we may use Pade approximations [4]. However, since the design and simulations will be in discrete time, it is reasonable to use the modified Z-transform that is applied to systems containing time delay elements [5]. Note that in the closed loop digital control system, there is a zoh (zero order hold) preceding the Gp(s). Therefore the modified Z-transform of (4) becomes

$$G_p(z) = \frac{z-1}{z} Z\left[\frac{z^{-t1} a_1 K_a e^{m_1 l s}}{s(s+a_2)}\right] + \frac{z-1}{z} Z\left[\frac{z^{-t2} a_2 K_2 e^{m_2}}{s(s+a_2)}\right]^{IS} \tag{5}$$

Which can be written

$$G_p(z) = \frac{zK_1 - zK_1 e^{-a_1} m_1 T + K_1 e^{-a_1 m_1 T} - K_1 e^{-a_1 T}}{z_1(z - e^{-a_1 T})} + \frac{zK_2 - zK_2 e^{-a_2 m_2 T} + K_2 e^{-a_2 m_2 T} - K_2 e^{a_2 T}}{z^{l_2}(z - e^{-a_2 T})}$$

Where

l: an integer,
m: a positive real number less than 1, *T* : sampling period,
$\tau_1 = l_1 T - m_1 T$ and $\tau_2 = l_2 T - m_2 T$.

If the nominal values of the parameters are used ($K1 =$-3, $K2 =$-7.3, $a_1 =$0.01, $a_2 =$0.006, $\tau_1 =$23s, $\tau_2 =$101s and T=10s) then the modified Z-transform of Gp(s) becomes

$$C_c = \frac{-0Z0ZBZZ^{10} - 0.10b34z^9 - 0.077855z^1 - 0.38375z^2 - 0.30583z - 0.037436}{z^{11} - 1.B466Z^{12} - 0.b5z14z^{11}} \quad (7)$$

Hence, the discrete time transfer function given by equation (7) is the mathematical model representing the patient (including the semi-closed circuit) in the simulation studies.

3. DESIGN OF THE FUZZY CONTROLLER

Fuzzy theory was first introduced by Zadeh in 1965 [6]. During the last two decades, Fuzzy Logic Control (FLC) has emerged as one of the most attractive and fruitful areas for research in the application of the fuzzy theory to the real engineering problems. FLC is actually a practical alternative to the conventional control methods for a variety of control applications since it provides a convenient method for implementing linear and non-linear controllers via the use of both heuristic and mathematical information [7].

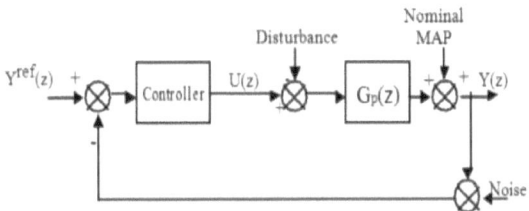

Figure 1: The Block Diagram of the discrete time closed loop control system

Figure 1 shows the block diagram of the discrete time closed loop control system used in simulation studies. Note that the controller should mimic the control actions of the anesthetist. In other words, the knowledge and experience of the anesthetist should be embedded into the controller. This fact suggests the use of a rule -based controller like fuzzy controllers. The nominal MAP value used in the simulation is 100. The controller output is limited with an anti-wind up integrator in order to avoid any overdose and improve the steady state performance. The upper and lower limits for the isoflurane concentration are chosen as 4% and 0%, i.e., the controller output u(t) saturates at the values of 4 and 0.

In this study, a Sugeno type fuzzy controller [7] is designed and its parameters are tuned such that the good output responses are obtained under the variations of parameters K1, K2, a1, a2, τ_1 and τ_2. The reason behind the test under parameter variations is the fact that these parameters change from one patient to another in the real applications. Therefore, in order to provide a robust control performance, the controller should be designed by taking account the parameter variations. The trail and error method is used in tuning of the controller parameters.

Figure 2 shows the input membership functions of the fuzzy controller. The first input is the error between the reference (or desired) MAP and the actual measured MAP. The second input is the change of this error.

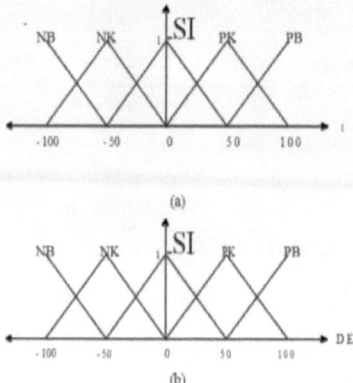

(a)

(b)

Figure 2: The inputs membership function

The rule base of the fuzzy controller is given in Table-1. The numeric values of the output variables are given in Table-2.

Table-1 : The Rule Base of the Fuzzy Controller

U E DE	NB	NK	SI	PK	PB
NB	D_1	D_2	D_3	D_4	D_5
NK	D_6	D_7	D_8	D_9	D_{10}
SI	D_{11}	D_{12}	D_{13}	D_{14}	D_{15}
PK	D_{16}	D_{17}	D_{18}	D_{19}	D_{20}
PB	D_{21}	D_{22}	D_{23}	D_{24}	D_{25}

Table-2: The Numeric Values Of The Output Variables

$D_1 = 4.4$	$D_2 = 4.175$	$D_3 = 3.95$	$D_4 = 3.725$	$D_5 = 3.5$
$D_6 = 2.375$	$D_7 = 2.15$	$D_8 = 1.925$	$D_9 = 1.7$	$D_{10} = 1.475$
$D_{11} = 0.35$	$D_{12} = 0.125$	$D_{13} = 0$	$D_{14} = 0.125$	$D_{15} = -0.35$
$D_{16} = -1.475$	$D_{17} = -1.7$	$D_{18} = -1.925$	$D_{19} = -2.15$	$D_{20} = -2.375$
$D_{21} = -3.5$	$D_{22} = -3.725$	$D_{23} = -3.95$	$D_{24} = -4.175$	$D_{25} = -4.4$

The values in Table-2 are determined by trial and error method as stated before. In the following section, simulation results showing the robust performance of the fuzzy controller under parameter variations are presented.

5. SIMULATION RESULTS

The closed loop system shown in Figure 1 is simulated and the robustness of the control performance is tested by changing the plant parameters.

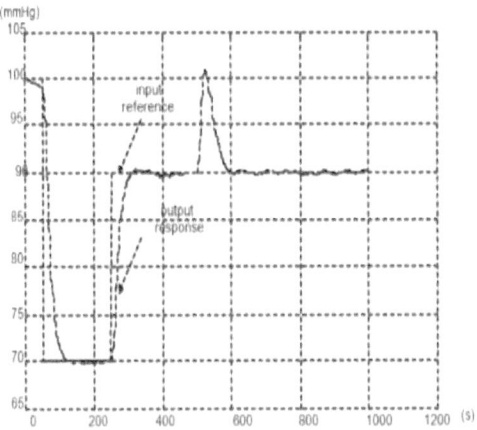

Figure 3: Output response for the nominal parameters

Figure 3 shows the output response to a step input profile for the nominal system parameters $K1 = -3$, $K2 = -7.3$, $a1 = 0.01$, $a2 = 0.006$, $\tau_1 = 23s$, $\tau_2 = 101s$. A noise signal and disturbance applied at $t = 500s$ are added as indicated in Figure 1. The output response to the same input step profile for $\tau_2 = 151.5s$ (50% increased) is shown in Figure 4.

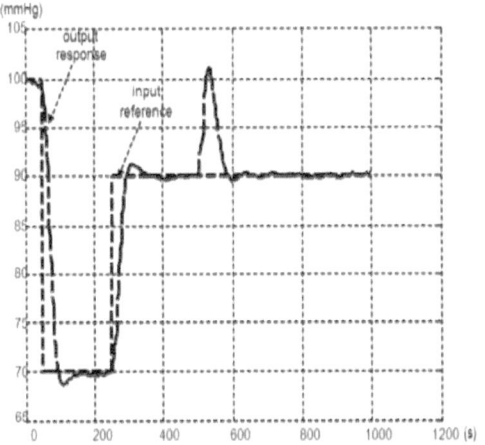

Figure 4: Output response for $\tau_2 = 151.5s$ (50% increased)

The output response for K2 = - 10.95 (50% decreased) is shown in Figure 5.

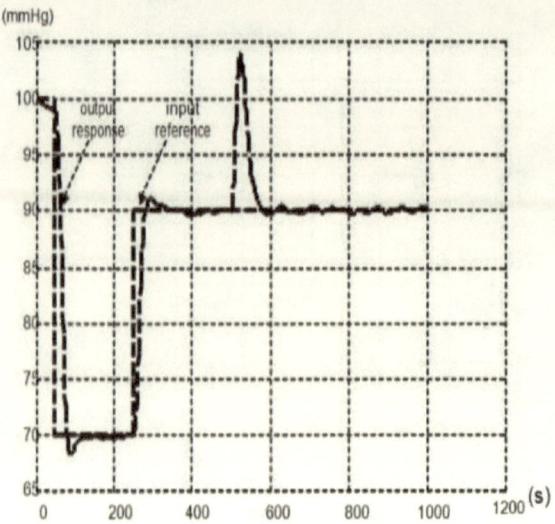

Figure 5: Output response for $K2 = -10.95$ (50% increased)

Figure 6 shows the output response for $a_2 = 0.018$ (200% increased).

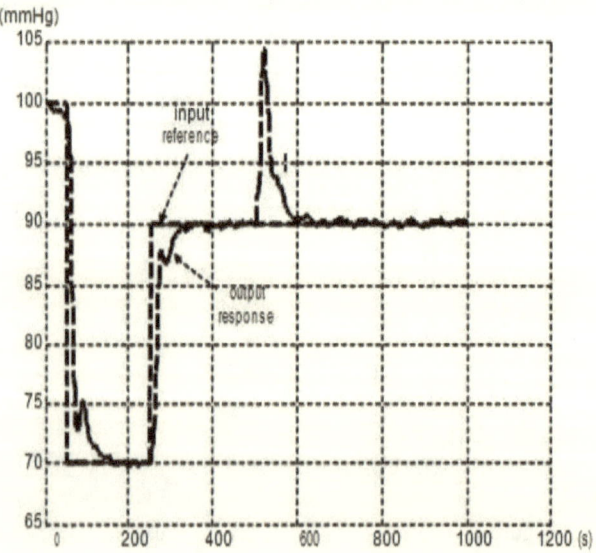

Figure 6: Output response for $a_2 = 0.018$ (200% increased)

Table-3: The maximum and minimum limits of the parameters which can be tolerated by the fuzzy controller

Parameters	Minimum values (decreasing)		Maximum values (increasing)	
	Percentage (%)	Values	Percentage (%)	Values
τ_1	100	0 s	200	69 s
τ_2	50	50.5 s	50	151.5 s
K_1	600	-21	50	-1.5
K_2	50	-10.95	30	-5.11
a_1	75	0.0025	200	0.03
a_2	30	0.0042	200	0.018

Table-3 shows the maximum and minimum limits of the parameters that can be compensated or tolerated by the fuzzy controller, i.e., even if the parameters change within the limits given in Table-3, the proposed fuzzy controller provides an acceptable (robust) control performance. Beyond this limits, the control performance is getting worse.

5. CONCLUSIONS

In this study, the control of the MAP of a patient under anesthesia is considered in order to prepare a basis for the real applications. A mathematical model representing the patient under anesthesia is derived and used in the simulation studies. A Sugeno type fuzzy controller is designed for the control of the MAP. The parameters of the controllers are determined by trial and error so that the robust control performances are obtained under plant parameter variations. The control performance of the proposed fuzzy controller is tested by changing the parameters of the mathematical model of the patient (including the semi-closed circuit). The simulation results showing the robust control performance of the fuzzy controller are presented in the paper. A table is given to show the maximum and minimum limits of the plant parameters that can be compensated by the fuzzy controller.

REFERENCES

[1]   R. Meier, J. Nieuwland, A. M. Zbinden and S. S. Hacisalihzade, "Fuzzy Logic Control of Blood Pressure during Anesthesia," *IEEE Control Systems Magazine*, vol. 12, no. 9, pp. 12-17, 1992.

[2]   S. Oshita, K. Nakakimura and T. Sakabe, "Hypertension Control during Anesthesia," *IEEE Engineering in Medicine and Biology*, vol. 13, no. 5, pp. 667-670, 1994.

[3]   D. G. Mason, and D. A. Linkens, "Hybrid Self-Organising Fuzzy Logic PID Controllers for Muscle Relaxant Anesthesia," in *Proc. UKACC International Conference on Control*, 1996, pp. 2-5.

[4]   G. F. Franklin, J. D. Powell and A. E. Naeini, *Feedback Control of Dynamic Systems*. Prentice-Hall, 2002.

[5]   G. F. Franklin, J. D. Powell and M. Workman, *Digital Control of Dynamic Systems*. Addison Wesley, 1998.

[6]   L. A. Zadeh, "Fuzzy sets", *Informat. Contr.*, vol. 8, pp. 338-353, 1965.

[7]   K. M. Passion, and S. Yurkovich, *Fuzzy Control*. Addison Wesley-Longman, 1997.

125

12

Development of A New Class of Natural Fiber Composite Material Using Human Hair and Coir Fibers

Saransh Tiwari [1], Sudhaker Dixit [2*] , Ved Kumar [2], Aakash Singh [4] Anod Kumar Singh [2]

[1] Indian Institute of Management Lucknow, IIM Road, Lucknow
[2] Department of Humanity and Applied Sciences, School of Management Sciences, Lucknow
[3] Preeminent Research Academy and Carrier Development Center (PRACD), Lucknow
[4] Dr. Shakuntala Mishra Rehabilitation University, Lucknow
*E-mail : dr.sudhkerdixit@gmail.com

ABSTRACT

The developed material is found to have certain exceptional properties which justify that the developed material might replace glass fiber composite and in certain cases can even replace carbon fibers. The results of the tests suggest thata maximum impact strength of 13.174 kJ/m²is obtained for the specimen with nomenclature IL3 which has human hair and coir in the ratio 3:2. From the results of water absorption test, it is found that the developed material is almost completely water resistant with maximum water absorption percentage of only 2.13% in WCH1 specimen. It is observed that the fiber orientation does not play a major role in water resistance, it is only the fiber content which plays an important role in water resistance. The maximum flexural strength obtain edis 36.270MPa for the specimen BL4 with hair and coir in the ratio 4:1. Further analysis of the results suggests that the properties of the developed natural fiber composite are influenced by parameters such as relative volume fraction of various fibers used, fiber orientation, and fiber aspect ratio among others. By repeated experimentation on specimen of different compositions and fiber orientations, the notable properties corresponding to each specimen are noted and the specimen with the most significant of each property has been marked to be utilized in future applications. Also, the developed material is eco-friendly and biodegradable in nature.

Keywords: *natural fiber composites; mechanical properties; water resistant test; bending test; impact test.*

1. INTRODUCTION

Fiber-reinforced composite (FRC) is a well-known composite material that consists of fibers which act as a reinforcing agent and a matrix phase to bind the fibers together, thereby producing a new material with combined properties of each of its constituents. A composite is a material that is formed by combining two or more materials which have different properties (physical, chemical, thermal, etc. properties) that, when combined, results in a formation of new material with certain characteristic properties which are different from that of individual constituents [1].Composite materials are distinguished by a number of desirable qualities, including a high strength-to-weight ratio, low weight, and design flexibility.

FRCs are high-performance fiber composites that are generated by cross-linking the cellulosic fiber molecules with the resins in the matrix using a patented molecular re-engineering technique.

This approach results in a product that has remarkable structural, physical, and chemical qualities. Using this technology of molecular re-engineering, selected physical, chemical, and structural properties of wood are cloned and vested in the FRC products. In addition, a variety of other critical attributes are incorporated into the process, which results in superior performance properties than those of contemporary wood [2]. This newly discovered material, in contrast to existing composites, has the potential to be recycled up to 20 times, which meant that scrap FRC could be utilized several times over. FRC materials can fail due to a variety of causes, including intra-laminar matrix cracking, de-lamination, longitudinal matrix splitting or separation, fiber-matrix de-bonding, fiber fracture, or fiber being dragged out. The composites are classified into three types as shown in Figure 1.

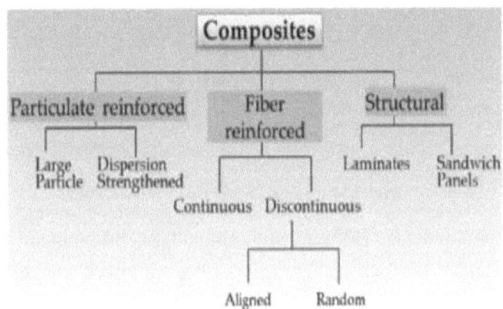

Figure 1: Types of composites[3]

1.1. Fiber reinforced composites

Fiber reinforced composites are classed as:

(i) Continuous (long fibers)
(ii) Discontinuous (short fibers).

When fibers are aligned in a particular direction, they provide very high strength but only along the direction in which they are aligned. The formed composite has a considerably weaker strength in other directions and therefore exhibits a significant anisotropic behavior. Figure 2 shows the structure of FRCs.

This anisotropy could be overcome by random orienting and aligning the fibers in all directions. This decreases the effective strength of the material but has the advantage of improving the formability of the material and reducing the cost[4]. For forming of materials with strength in only one direction, aligned fibers are used while for having strength in all directions randomly aligned fibers are used.

Composites that require fibers being aligned in only one principal direction can be formed using either continuous or dispersed/discontinuous fibers while composites with randomly aligned fibers are mostly made with discontinuous fibers (Figure 3).

Composites may be considered to be the tailor-made materials in which there are various parameters, other than the properties of the fibers and matrix, which could be altered to meet the design and application requirements of the manufacturer.

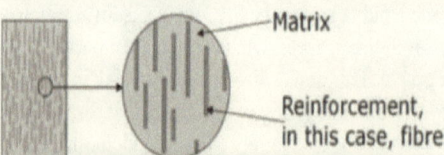

Figure 2: Structure of fiber reinforced composites.

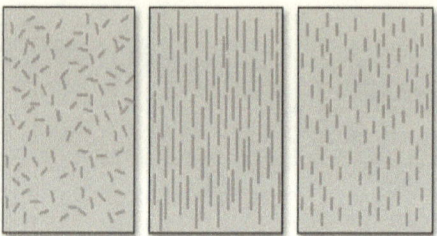

(a) Continuous aligned.(b) Discontinuous aligned.(c) Discontinuous random.
Figure 3: Different types of alignment in fiber reinforced composites.

1.2 Structural and Chemical Constituents of Human Hair and Coir Fiber

Hair is a protein filament that develops from dermal follicles. Hair is an essential biomaterial comprised mostly of protein, specifically alpha-keratin. The entire chemical makeup of hair consists of 45% carbon, 28% oxygen, 15% nitrogen, 7% hydrogen, and 5% sulphur.

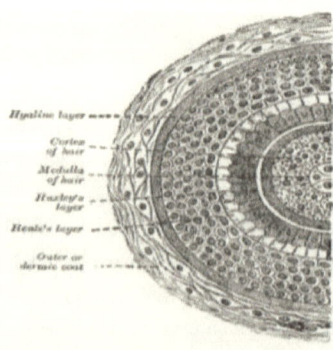

Figure 4: Cross-section of human hair [7].

Coconut produces coir fibers between its tough inner shell and its outer cover. Each fiber cell is slender, hollow, and has thick walls holding cellulose. As a coating of lignin is formed over young coir fibers, they eventually harden and turn yellow (Figure 6). A typical fiber cell is around 1 mm (0.04 inches) long and 12 to 22 m (0.0004 to 0.0008 inches) in diameter (transverse direction). The average length of a fiber is between 10 and 30 cm (4 and 12 inches). The two types of coir are brown and white. The brown coir derived from mature coconuts is robust, dense, and possesses a good abrasion resistance [8]. They are predominantly employed for sacking, matting, and brushes. Mature brown coir fibers have more lignin and less cellulose than other fibers, such as cotton, and are consequently more rigid and less flexible. White coir fibers are extracted from unripe coconuts and are typically smooth, fine, and weak. Coir fibers are one of the few natural fibers that are salt-water resistant [9]. White coir may be manufactured using either fresh or salt water.

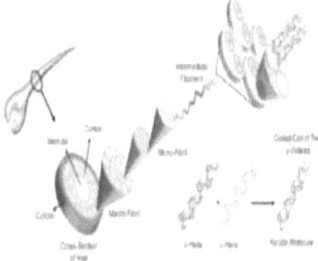

Figure 5 α: Keratin: a molecule with a helical structure [10].

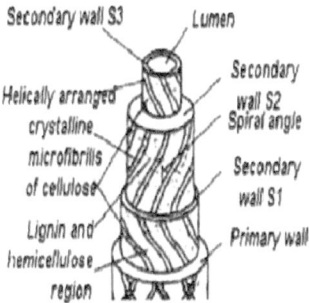

Figure 6: Structure of coir fiber [11].

In this study, we aim to develop a new composite material using natural fibers (human hair and coir) and study its mechanical properties. By proper study and efficient utilization of human hair and coir as fibers in different compositions together with polyester resin as matrix, various specimens were prepared using hand layup technique and were send for testing at the CIPET lab. After their respective testing, it was found that the properties of the developed material are such that it may replace several existing materials.

Through an exhaustive literature review, it has been observed that the bending and impact strength properties of the human hair and coir fibers reinforced composite has not been investigated precisely prior our work. In this paper, we have also attempted to explore the water-resistant property and the maximum permissible break load that the developed composite can bear.

The rest of the paper is structured as follows: Section 2 describes the methodology employed for the fabrication of the composite material, Section 3 highlights the geometry of the developed specimens and the orientation of fibers, Section 4 presents the nomenclature of the specimens, followed by Section 5 which highlights the properties of the materials used to fabricate the composite material, Section 6 describes the tests performed on the specimens, Section 7 reports the results of this study, followed by Section 8 which draws the conclusion.

2. METHODOLOGY: FABRICATION AND SYNTHESIS OF THE COMPOSITE MATERIAL

Composite synthesis includes wetting, blending or saturating the reinforcement with the matrix, and then allowing the matrix to bind together (either with thermal action or chemical reaction) into a rigid structure [12, 13]. In the present work, the composite material samples are developed using the hand lay-up technique.

Fiber extraction

Firstly, the fiber is extracted from their natural sources by means of simple bare hand techniques or by means of simple tools. Coir is extracted from coconut and hair is extracted from human body.

2.1 Preparation of fiber

Following procedure is employed for preparation of fiber:

(i) Firstly, from the extracted coir a fiber of appropriate size and length is taken and rubbing action is performed by means of sandpaper.

(ii) After rubbing action small numbers of fibers are selected in groups for strand preparation.

2.2 Preparation of mould

A mould is prepared by carving out a cavity of required dimension in a wooden block. A small allowance is given in mould in order to overcome the filing action.

2.3 Orientation of fibers

Orientation is termed as the alignment of fibers in the mould along with resin mix. The properties of fabricated composites are also defined by the orientations of fibers (Figure 7). The fiber strands are selected and weighed in required amount, then these fibers are aligned as per demand in mould and in order to withstand that alignment against melding action they are fixed in some of the places by means of strong adhesive.

Figure 7: Theorientation of fibers

2.4. Melding action

Within a mould, the reinforcing and matrix materials are combined, compacted, and cured (processed) to accomplish a melding event. Small amount of hardener MEKP and cobalt catalyst is added for fast preparation of composite.

2.4 Removal of specimen and cleaning

The specimen is removed from the mould and then cleaning is done by filing process. Filing is not only performed for smooth surface finish but it is also used for obtaining dimension accuracy.

2.5 Testing

Obtained specimens are tested in the lab and resulting data is processed for further analysis. In order to select the best suitable composition for composite preparation.

Figure 8 depicts the sequence of operations adopted in synthesis the composite.

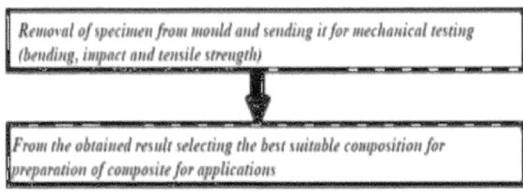

Figure 8: Flowchart for composite fabrication

3. ORIENTATION OF THE FIBERS AND THE GEOMETRY OF THE DEVELOPED COMPOSITE SPECIMENS

Estimated weight of prepared composite = 20 grams.

Polyester resin matrix: Ratio of Fiber Reinforcement = 9:1.

The different compositions of hair fiber and Coconut fiber used for composite preparation are reported in Table 1.

Table-1: Relative Percentage of Fibers used in Composites

S.No.	Human hair fiber (%)	Coir fiber (%)
1	80	20
2	60	40
3	40	60
4	20	80

The pattern is prepared in a mould of dimension (LxBxH) cm³. Each of these pattern have its own characteristic mechanical property which would be justified upon subjecting them to different testing.

3.1. *Longitudinal Loading Pattern*

Figure 9: Longitudinal loading pattern

In this pattern, fibers of coconut are placed parallel to one another and the human hair is placed in a dispersed manner and the fibers are bonded using polyester matrix (Figure 9). The loading is applied axially in the direction of coconut fiber alignment.

3.2. *Transverse Loading Pattern*

Figure 10: Transverse loading pattern

In this pattern, fibers of coconut are placed parallel to one another and the human hair is placed in a dispersed manner and the fibers are bonded using polyester matrix (Figure 10). The loading is applied perpendicular to the direction of coconut fiber alignment.

3.3. Cross-hatched Loading Pattern

Figure 11: Cross-hatched pattern

In this pattern, fibers of coconut are placed in a cross-linking manner like a wire mesh and human hair is placed in a dispersed manner and the fibers are bonded using the polyester matrix (Figure 11). Loading is applied in a horizontal direction perpendicular to the sample face.

4. THE NOMENCLATURE OF SPECIMENS

The following system of nomenclature has been used for the effective recognition of the specimens for the various experiments:

(i) On the basis of tests being performed:

I = impact test, B = bending test, and W= water absorption

(ii) On the basis of orientation of fibers in the specimen:

L = longitudinal, T = transverse, and CH = cross-hatched

(iii) On the basis of relative composition of hair and coir in the specimen – The samples are taken in the ratio of the composition of the hair to coir as follows: Sample 1: hair:coir = 1:4, Sample 2: hair:coir = 2:3, Sample 3: hair:coir = 3:2 and Sample 4: hair:coir = 4:1.

5. MATERIALS USED

The raw materials used in this study are as follows:

(i) Human Hair

(ii) Coir

(iii) Polyester resin

(iv) Hardener MEKP

(v) Catalyst cobalt

Human hairs have been collected from the nearby waste dump of barber's shop. Coir fibers were obtained from the fruit-seller dustbin. Polyester resin, MEKP and Cobalt were bought from online seller.

5.1 Human Hair

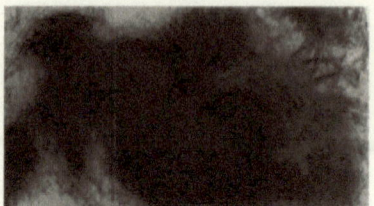

Figure 12: Human Hair

5.2. Coir Fibers

Coir (Young's modulus between 3 and 4 GPa) is a natural fiber produced from coconut husks. Coir is a fibrous material created between a coconut's interior shell and its outer husk (Figure 13). The individual fiber cells are thin, dense, and hollow, and are mostly composed of cellulose. Its properties are:

(i) Coir fibers are multi-cellular, hard, very coarse in nature

(ii) They are agro-renewable, biodegradable and a good blend of extensibility, strength, length, moisture regain, and have high durability against solar radiations [8].

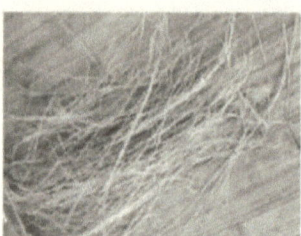

Figure 13: Coir fibers

5.3. Polyesterresin

Figure 14: Polyester Resin

Polyester is an unsaturated synthetic resin which is produced by a reaction between polyhydric alcohols and dibasic organic acids[15]. Polyester can be used as a matrix, which is usually a viscous material that solidifies and hardens to give suitable shape and size to the composite product and to protect the fibers from deterioration (Figure14). Some of its properties are as follows:

(i) Adequate resistant to weathering, tearing and ageing with passage of time.

(ii) They are moderately temperature resistant, can survive well upto 80^0C.

(iii) They undergo relatively low shrinkage of upto 4-8% during curing.

(iv) They have a low thermal expansion coefficient which ranges between $(100-200) \times 10^{-6} K^{-1}$.

5.1 Hardener - MEKP

In industry and by hobbyists, dilute solutions of 30 to 60% MEKP are employed as a catalyst to induce cross-linking between unsaturated polyester resins used in fiber glass and castings. To lessen shock sensitivity, MEKP is typically dissolved in cyclohexane peroxide, dimethyl phthalate, or diallyl phthalate. Additionally, benzoyl peroxide can be utilized for the same purpose.

MEKP is a severe skin irritant and may cause progressive corrosive damage and in certain cases blindness[16]. Therefore, it should be carefully used.

5.2 Catalyst - Cobalt

Cobalt catalyst is a highly active reagent, extensively used in the efficient and selective formation of pharmaceuticals, natural products and other new materials. Cobalt catalyst has a higher reactivity for numerous C-C bond formations. Cobalt catalyst with cobalt salts show good functional group tolerance, mild reaction conditions and high chemo-selectivity in comparison with nickel and palladium [17]. They are the most commonly used catalyst for metal-catalyzed cross-coupling reactions.

6. TESTING AND ANALYSIS PROCEDURE

The following tests are conducted on the fabricated composite specimens.

6.1 Bending Test

Bend testing systems accurately and reliably measure the flexural properties of materials. The transverse bending test is often done using this machine in which a specimen of given dimensions having either a circular, square or a rectangular cross-section is bent by means of a bending moment using a three-point or four-point flexural test technique until fracture or yielding of material is produced (Figure 15).The flexural strength is represented by the maximum stress experienced within the material at its moment of yield.

Figure 15 : Four-point bending machine

6.2 Water Absorption Test

Moisture absorption test of composite involves placing of material in a moisture prone environment and then monitoring the changes in its weight with passage of time. Initially, the weight of the specimen is recorded and then after regular time intervals its weight is measured. The relative change in the weight of specimen gives information about its water-resistant behavior (Figure 16).

Figure 16: Water absorption testing machine

6.3 Impact Test

Impact tests provide information about the toughness of material. The toughness of a material is defined as its ability to absorb energy upto the fracture point.

ASTM standardizes the Izod impact test for determining the impact resistance of materials. It is composed of a pivoting arm that is lifted to a certain height (constant potential energy) and then released (Figure 17). The arm swings down to strike a sample with a notch, fracturing the sample. The specimen's toughness or energy absorption is determined by measuring the height to which the arm is elevated after striking the specimen. Typically, a specimen is notched to compute the hardness and notch sensitivity as necessary (Figure 18).

Figure 17: Izod Impact Testing Machine

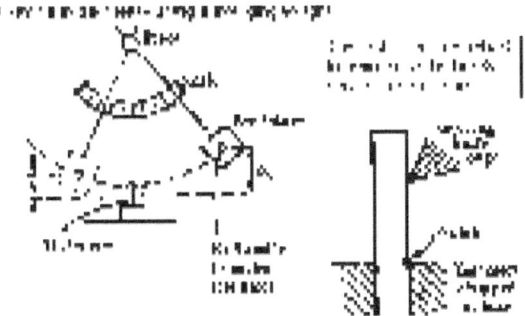

Figure 18 : Impact testing procedure

7. RESULTS

This section illustrates the results acquired by testing composite specimens and their appropriate analysis.

7.1 Bending test results –(machine used: ASTM D 790)

(i) BL1: Figure 19 reports the result for BL1 specimen.

Dimension: 13x125x5

Resin: Fiber = 9:1

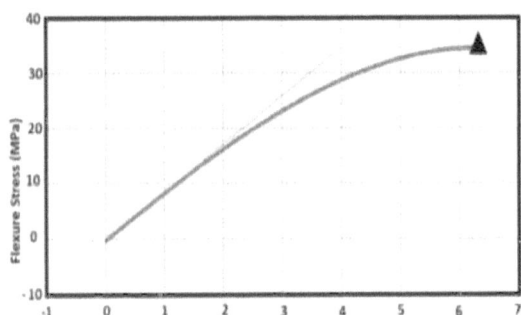

Figure 19: Graph between flexure stress and flexure strain for BL1 specimen

Coir: Hair = 4:1

(ii) BL2 : Figure 20 reports the result for BL2 specimen.

Dimension: 13x125x5

Resin: Fiber = 9:1

Coir: Hair =3:2

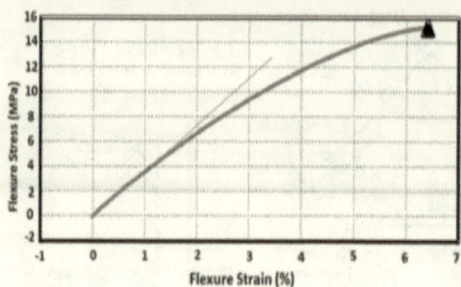

Figure 20: Graph between flexure stress and flexure strain for BL2 specimen.

(iii) **BL3 :** Figure 21 reports the result for BL3 specimen.
Dimension: 13x125x5
Resin: Fiber = 9:1
Coir: Hair=2:3

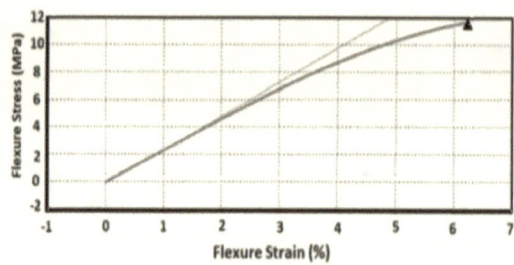

Figure 21 : Graph between flexure stress and flexure strain for BL3 specimen

(iv) **BL4 :** Figure 22 reports the result for BL4 specimen.
Dimension: 13x125x5
Resin: Fiber = 9:1
Coir: Hair=1:4

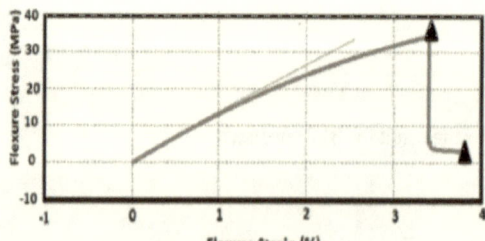

Figure 22: Graph between flexure stress and flexure strain for BL4 specimen.

(v)   **BT1 :** Figure 23 reports the result for BT1 specimen.

Dimension: 13 x 125 x 5

Resin: Fiber = 9:1

Coir: Hair = 4:1

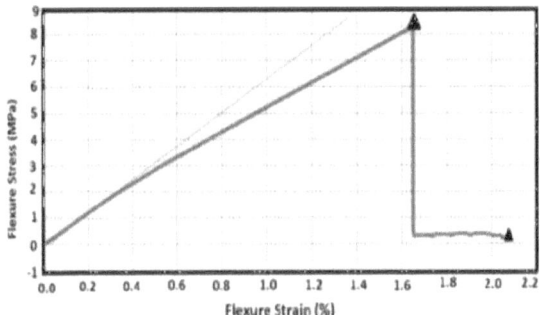

Figure 23 : Graph between flexure stress and flexure strain for BT1 specimen

(vi)   **BT2 :** Figure 24 reports the result for BT2 specimen.

Dimension: 13x125x5

Resin: Fiber = 9:1

Coir: Hair = 3:2

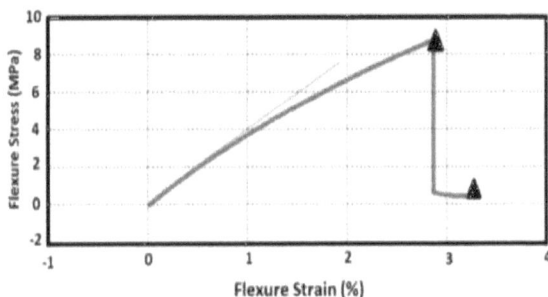

Figure 24: Graph between flexure stress and flexure strain for BT2 specimen.

(vii)   **BT3 :** Figure 25 reports the result for BT3 specimen.

Dimension: 13x125x5

Resin: Fiber = 9:1

Coir: Hair = 2:3

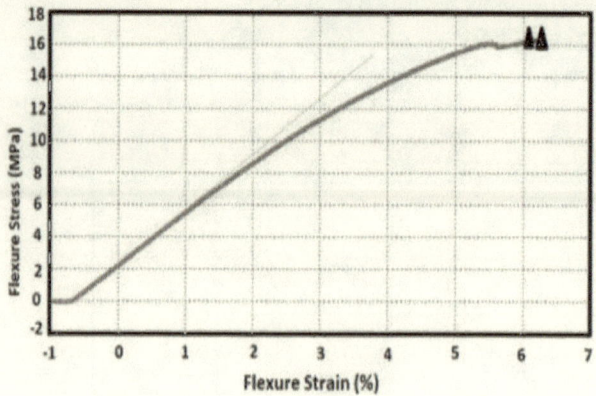

Figure 25 : Graph between flexure stress and flexure strain for BT3 specimen

(viii) **BT4 :** Figure 26 reports the result for BT4 specimen.
Dimension: 13x125x5
Resin: Fiber = 9:1
Coir: Hair =1:4

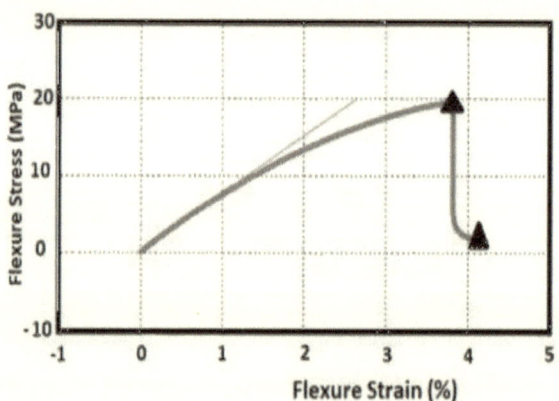

Figure 26 : Graph between flexure stress and flexure strain for BT4 specimen

(ix) **BCH1 :** Figure 27 reports the result for BCH1 specimen.
Dimension: 13x125x5
Resin: Fiber = 9:1
Coir: Hair =4:1

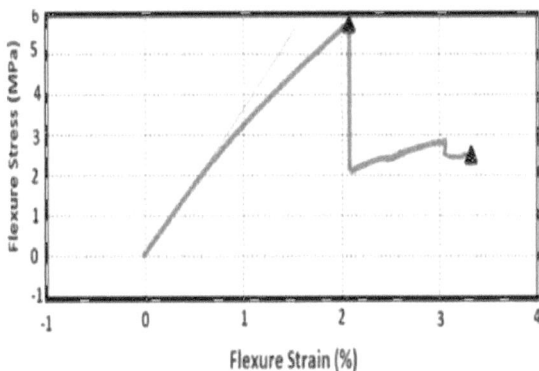

Figure 27: Graph between flexure stress and flexure strain for BCH1 specimen

(x)   **BCH2 :** Figure 28 reports the result for BCH2 specimen.

Dimension: 13x125x5

Resin: Fiber = 9:1

Coir: Hair =2:3

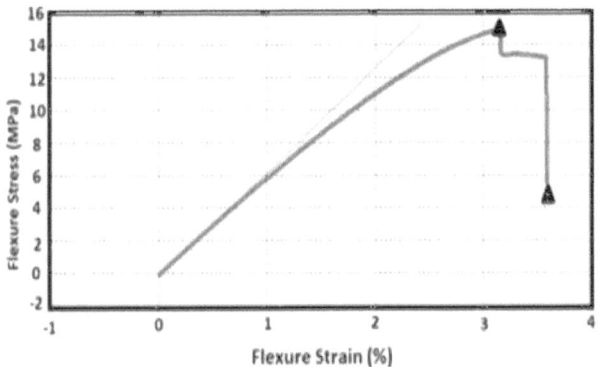

Figure 28: Graph between flexure stress and flexure strain for BCH2 specimen

(xi)   **BCH3 :** Figure 29 reports the result for BCH3 specimen.

Dimension: 13x125x5

Resin: Fiber = 9:1

Coir: Hair =2:3

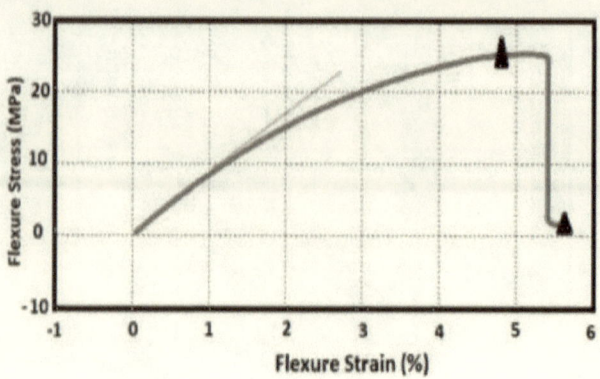

Figure 29 : Graph between flexure stress and flexure strain for BCH3 specimen

(xii)   **BCH4 :** Figure30 reports the result for BCH4 specimen.
         Dimension: 13x125x5
         Resin: Fiber = 9:1
         Coir: Hair = 1:4

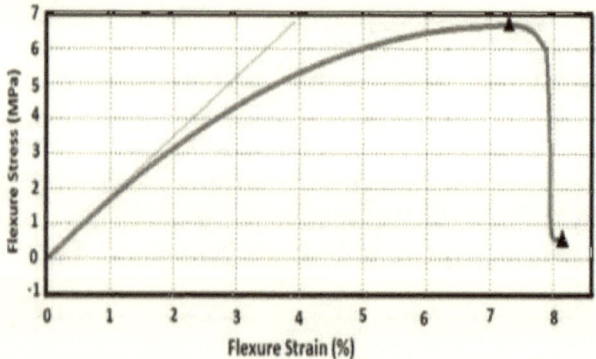

Figure 30 : Graph between flexure stress and flexure strain for BCH4 specimen

Table-2 reports the detailed results of the bending test for all composite specimens.

Table-2: Results of bending test

S. No.	Nomenclature	Support span (in mm)	Speed (mm/ min)	Maximum flexure load (N)	Flexure stress at max. flexure load (MPa)	Flexure load @break (N)	Flexure strength @ break (MPa)	Flexure strain at break (%)	Width (mm)	Thickness (mm)
1.	BL1	100.0	3.0	126.5	35.2	125.8	35.0	6.2	11.0	7.0
2.	BL2	100.0	3.0	53.4	14.8	52.7	14.6	6.2	11.0	7.0
3.	BL3	100.0	3.0	43.0	11.9	42.3	11.7	6.2	11.0	7.0
4.	BL4	100.0	3.0	130.3	36.2	16.7	4.6	3.9	11.0	7.0
5.	BT1	100.0	3.0	83.8	9.0	1.1	0.1	3.1	11.5	11.0
6.	BT2	100.0	3.0	76.5	8.2	2.7	0.2	2.1	11.5	11.0
7.	BT3	100.0	3.0	59.9	16.6	59.4	16.5	6.2	11.0	7.0
8.	BT4	100.0	3.0	73.6	20.4	6.6	1.8	4.0	11.0	7.0
9.	BCH1	100.0	3.0	66.9	5.9	29.9	2.6	3.4	11.0	11.0
10.	BCH2	100.0	3.0	56.9	15.8	16.8	4.6	3.6	11.0	7.0
11.	BCH3	100.0	3.0	85.6	23.8	6.3	1.7	5.6	11.0	7.0
12.	BCH4	100.0	3.0	63.4	6.8	3.3	0.3	8.1	11.5	11.0

7.2 Impact test results - (machine used:ASTM D 256)

Impact test was conducted on 12 specimens of different compositions and fiber orientation of specimen size (13x62x5). Three fiber orientations were selected and corresponding to each orientation 4 specimen of different compositions were prepared for testing and analysis purpose. From the results obtained from the experiment, corresponding graphs were plotted. The obtained results with their corresponding graphs are-

(i) Longitudinal fiber orientation (refer to Table 3 and Figure 31).

Table-3: Impact test result for longitudinal fiber orientation.

S.No.	Specimen	Impact strength (kJ/m^2)
1	IL1	11.9688
2	IL2	5.4237
3	IL3	13.1744
4	IL4	6.4302

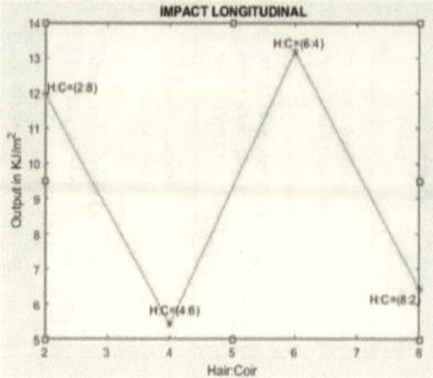

Figure 31 : Graphical analysis of impact test result for longitudinal fiber orientation.

(ii) Transverse fiber orientation (refer to Table 4 and Figure 32).

Table-4: Impact test result for transverse fiber orientation.

S.No.	Specimen	Impact strength (kJ/m²)
1	IT1	3.7342
2	IT2	2.7600
3	IT3	4.5364
4	IT4	10.3982

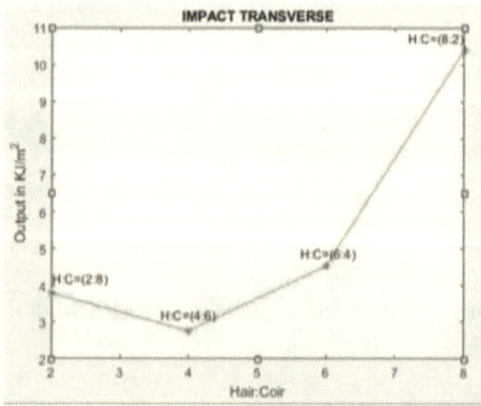

Figure 32: Graphical analysis of impact test result for transverse fiber orientation

(iii) Cross-hatched fiber orientation (refer to Table 5 and Figure 33).

Table-5: Impact test result for cross-hatched fiber orientation.

S.No.	Specimen	Impact strength (kJ/m^2)
1	ICH1	3.4513
2	ICH2	4.7484
3	ICH3	2.6226
4	ICH4	2.1098

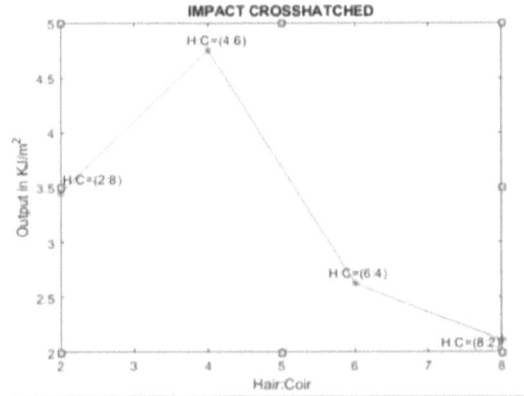

Figure 33: Graphical analysis of impact test result for cross-hatched fiber orientation.

Combining all the above obtained data, it can be concluded that the specimens with longitudinal fiber orientation yielded better impact strength compared to transverse and cross-hatched fiber orientations. Cross-hatched fiber oriented composite specimen are not much impact resistant.

Maximum obtained Impact Strength- 13.1744 KJ/m^2 in IL^3 specimen.

Minimum obtained Impact Strength- 2.1098KJ/m^2 in ICH^4 specimen.

7.3 Water absorptiontestresults – (machine used:ASTM D 570)

Water absorption test was conducted on eight specimens of different compositions and fiber orientation on specimen of size (50x50x5). Two fiber orientations were selected and corresponding to each orientation 4 specimen of different compositions were prepared for testing and analysis purpose. From the results obtained from the experiment, corresponding graphs were plotted. The obtained results with their corresponding graphs are-

(i) Longitudinal/Transverse fiber orientation (refer to Table 6 and Figure 34)

Table-6: Water absorption result for longitudinal/transverse orientation.

S. No.	Specimen	Weight of specimen before testing W_1 (in grams)	Weight of specimen after testing W_2 (in grams)	% Water Absorption = $\dfrac{W_2 - W_1}{W_1} \times 100$ (in %)
1	WL1	19.3280	19.7750	0.98
2	WL2	17.1229	17.3102	1.09
3	WL3	25.7233	25.9405	0.84
4	WL4	24.0451	24.2450	0.83

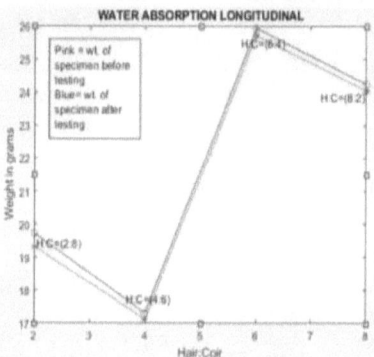

Figure 34: Graphical analysis of water absorption test results for longitudinal/transverse fiber orientation.

(ii) Cross-Hatched fiber orientation (refer to Table 7 and Figure 35)

Table-7: Water absorption result for cross-hatched orientation.

S. No.	Specimen	Weight of specimen before testing W_1 (in grams)	Weight of specimen after testing W_2 (in grams)	% Water Absorption = $\dfrac{W_2 \quad W_1}{W_1} \times 100$ (in %)
1	WCH1	22.1498	22.6872	2.43
2	WCH2	23.1233	23.3992	1.19
3	WCH3	17.6850	17.9207	1.33
4	WCH4	20.5783	20.7744	0.95

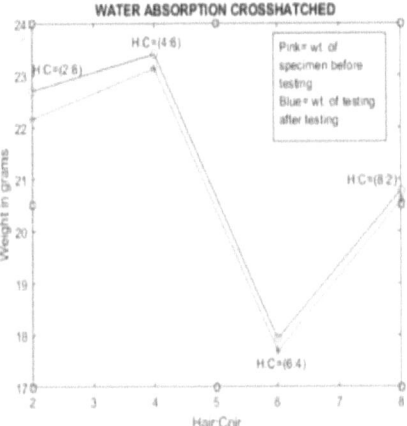

Figure 35: Graphical analysis of water absorption test results for cross-hatched fiber orientation. From the obtained experimental results, it can be successfully concluded that almost all the specimen are moisture resistant.

7.4 Brief comparison among hair-coir, glass fiber and carbon fiber composites

Table 8 illustrates the basic differences among the hair-coir, glass fiber and carbon fiber composites [18, 19].

Table-8: Comparison among hair-coir, glass fiber, and carbon fiber composite.

S. No.	Parameter	Hair-coir composite	Glass fiber composite	Carbon fiber composite
1.	Impact strength (kJ/m²)	13.1744	11.6	45 but for some grades 13.4
2.	Flexural strength (MPa)	36.27	28.25	500
3.	Water absorption (%)	2.13	5-9	0.5
4.	Average value of Young'smodulus (GPa)	1.194	2.01	250
5.	Approximate cost of fibers	inexpensive	180/kg	3000/piece
6.	Applications	Packaging material, etc.	Electrical insulation, etc.	Aeronautical industry, etc.
7.	Special advantage	Biodegradable, waste reduction	-	-

5. CONCLUSION

The developed composite material owing to its desirable properties can be utilized for essential local and commercial applications such as development of packaging material. The developed material is found to have exceptional properties which make is suitable to be utilized in the following future applications: (i) Development of packaging materials of high impact strength. (ii) Development of blades of low power generating capacity wind mills.The creation and evaluation of prototypes made from the hair-coir reinforced composite material for the aforementioned applications is a possible area of focus for research to be conducted in the future.

REFERENCES

[1]   P. S, S. KM, N. K, S. S: J. Mater. Sci. Eng. 06 03 (2017)

[2]   O. Faruk, A. K. Bledzki, H.-P. Fink, M. Sain: Macromol. Mater. Eng. 299 1(2014)

[3]   R. Paul, L. Dai: Compos. Interfaces 25 7 (2018)

[4]   X. Li, L. G. Tabil, S. Panigrahi: J. Polym. Environ. 15 1 (2007)

[5]   Z. Hu, G. Li, H. Xie, T. Hua, P. Chen, F. Huang: Proc.SPIE 7522 75222 (2010)

[6]   M. Zimmerley, C.-Y. Lin, D. C. Oertel, J. M. Marsh, J. L. Ward, E. O. Potma: J. Biomed. Opt. 14 4(2009)

[7]   H. Gray, C. M. Goss: Am. J. Phys. Med. Rehabil. 53 6 (1974)

[8]   S. Biswas, Q. Ahsan, A. Cenna, M. Hasan, A. Hassan: Fibers Polym. 14 10 (2013)

[9]   M. Ali: J. Civ. Eng. Constr. Technol. 2 6 (2010).

[10]   H. Lee: Tissue Eng. Regen. Med. 11 4 (2014)

[11]   S. Jayabal, S. Sathiya Murthy, K. T. Loganathan, and S. Kalyanasundaram: Bull. Mater. Sci. 35 4(2012)

[12]   S. Choudhry, B. Pandey: Int. J. Mech. Ind. Eng. 1098 2231 (2013)

[13]   M. P. N. E. Naveen: IOSR J. Mech. Civ. Eng. 2 3 (2012)

[14]   S. P. Selvan, V. Jaiganesh, K. Selvakumar: Int. Conf. Adv. Des. Manuf. 12 3 (2014)

[15]   G. Gunduz, D. Erol, N. Akkas: J. Compos. Mater. 39 17 (2005)

[16]   F. T. Fraunfelder, D. J. Coster, R. Drew, F. W. Fraunfelder: Am. J. Ophthalmol. 110 6 (1990)

[17]   A. Choya, B. de Rivas, J. R. González-Velasco, J. I. Gutiérrez-Ortiz, R. López-Fonseca: Appl. Catal. A Gen. 591 117381 (2020)

[18]   T. P. Sathishkumar, S. Satheeshkumar, J. Naveen: J. Reinf. Plast. Compos. 33 13 (2014)

[19]   N. van de Werken, H. Tekinalp, P. Khanbolouki, S. Ozcan, A. Williams, M. Tehrani: Addit. Manuf. 31 100962 (2020)

13

Path Optimisation in Manet Using Nemo

Amod Kumar Pandey[1], Hemant Kumar Singh[2], Sunit Kumar Mishra[3],
Himanshu Kumar Shukla[4] and Krishna Nand Mishra[5]

[1]Department of Electronics & Communication Engg., School of Management Sciences Lucknow
[2-3]Department of Computer Science & Engg., School of Management Sciences Lucknow
[4]Department of Computer Science & Engg., FOET, Lucknow University,
[5]Department of Computer Science & Engg., KMCL University, Lucknow.

ABSTRACT

This paper gives a brief description of connectivity between two MANETs using the concept of NEMO. Various wireless routing protocols are used which have their own advantages and disadvantages. This solution deals with optimising the path during conditions when the mobile network goes to a different location from the present location. This method reduces the data loss and also maintains the time of transfer. We have also tried to reduce the delay generated and an optimal path for the transfer of data. To avoid data loss we use binding registration in the peer to peer connectivity which can cause substantial time delay. We are basically dealing with reduction through route optimisation.

Keywords: *MANETs, NEMO, Route optimisation, Routing protocols.*

1. INTRODUCTION

In today's scenario, wireless technology played an important role. IETF (Internet Engineering Task Force) designed Mobile IP because in current internet status a user cannot move from one place to another place without breaking the IP communication i.e. they cannot change the access router. So Mobile IP was designed to provide mobility to the internet.

A Mobile node has two IP addresses i.e. HoA (Home address) and CoA (Care of address). The first HoA identifies the identity of the mobile node and second address specifies the current location of the Mobile node. A Mobile node is accessed by HoA but it changes the CoA with respect of the movement of the mobile node. In the home network a router called as Home Agent is used, which is placed at the Mobile node's (MN) home network for binding HoA and CoA address.

Many limitations arise in Mobile IP due to conditional sending of data through HA. It becomes the bottleneck for the whole system but when packets are routed through the specific HA it reduces the communication performance, increases the delay and infrastructure load. A single HA serving several MN's connection will lead to its failure. Mobile IPv6 easily solves the problem of route optimization by allowing several MN's to communicate with their peers directly by exploiting special IPv6 headers. The different version of NEMO protocol i.e. v4 and v6 which mainly provides mobility to network rather than the node and it doesn't support route optimization concept even in IPv6. In result, route optimization issue is an important task to carry out in the current internet status i.e. in IPv4 and even in IPv6 (in future).

NEMO i.e. Network that Moves is mainly seen as a Mobile Community Network. From the internet infrastructure point of view, a community network is set of nodes which are located in the same geographical area. The nodes are equipped with at least one wireless interface and will share its information directly by using an ad hoc protocol. With the connection point of view, nodes belonging to the community networks share the common point of attachment which acts as the NEMO's mobile router. This router has mainly two interfaces i.e. external interface and internal interface. The external interface equipped with long range wireless interface that is intended to attach to the internet and the internal interface provides connectivity to the internal nodes of the community networks.

To solve the problem of route optimization, different solution has been proposed, multiple Home Agent (HA) concepts is being deployed in different ASes (Autonomous System). For this a mobile node will choose the best HA (Home Agent) according to its topological position and in result it will choose the optimal path by reducing the communication delay of the path towards its peers.

In this paper, a scalable architecture is proposed which will solve the route optimization problem in Mobile IP and in NEMO.

2. SYSTEM ARCHITECTURE

The main goal of the above architecture is to reduce the communication delay of the MN and load at the fHAs. In the above, when a Mobile IP or NEMO client changes its point of attachment to the internet then it will establish a new tunnel with its HA for communication. When the MN detects that the current path assigned to it for communication has unacceptable performance then it queries the original HA for a close one.

The architecture is flexible. Several Has are deployed throughout the Internet and it has four differentiated phases. HAs organizes themselves in to a peer-peer network to store the information regarding their address and there topological information. The MN always bound to a HA belonging to this P2P network. In this when a MN detects that the RTT to its current HA is greater than the threshold then it will triggers the fHA discovery phase and start queries the P2P network for closer HA . Once the MN has the IP address of closer HA then it sends a Binding Update message (registration message) and obtains a new HoA.

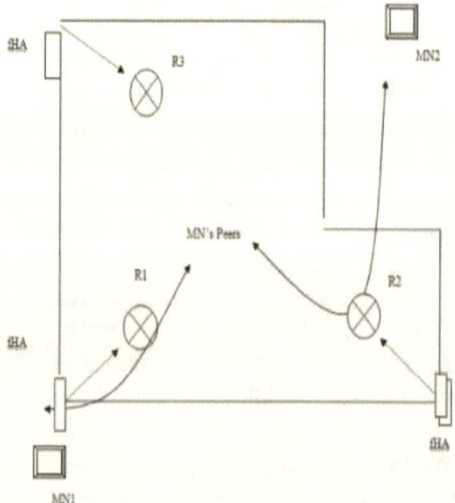

Figure 1: Architecture of fHN

2.1. P2P Setup Phase

This section will describe how the P2P network is created. The main concept of the P2P network is to store the location of fHAs i.e. AS number and their IP addresses which is used by MNs to locate the closer fHA to its topological position. fHAs organize themselves to form a structured P2P network. When a new fHA joins the fP2P–HN it chooses an identifier (Peer-ID). The fHA's position in the ring is determined by its Peer-ID: the fHA is placed between the two overlay nodes with the immediately higher and lower Peer-ID to its own id. Each overlay node has direct references to its two neighbours and also to other overlay nodes (crossing the ring) thus making the routing within the fP2P–HN faster. These nodes are named fingers. Each overlay node uses these fingers to create its fP2P–HN routing table. Finally, each fHA must register its AS number within the fP2P–HN. The successor stores an entry with all this information.

2.2. fHA Discovery Phase

When a mobile node (MN) connected to its current HA found that it is taking longer RTT above a given threshold, then after that it will initiate the discovery phase for a closer HA.

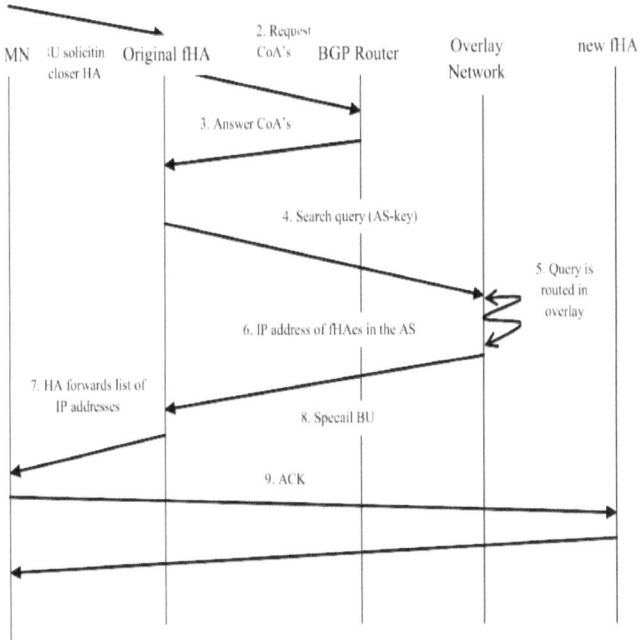

Figure 2: fHA Discovery phase

2.3. Data Packet Forwarding Phase

If the MN is connected to the fHA's autonomous system (AS), then it follow the same procedure as just in Mobile IP or in NEMO. But when the MN is attached to the foreign AS, then the MN should forward the packets through its HA. However, when the HA is an fHA, the MN encapsulate its data packets towards the BRs. Since the fHA has previously configured a new tunnel (Home

Address\   32 Tunnel) in the BRs, packets sent by the MNs are automatically de-capsulated and forwarded towards the packet's destination address (the MN's peer address). If the exit point of the MN's peer address is another BR, then the packet traverses the network as a transit packet. When the packets addressed towards the MN's HoA then, they will reach the fHA's AS. The BRs have learned the location (CoA) of the MN through IBGP and will automatically encapsulate and forward the packet directly towards the MN.

3. RESULT

In this paper, NS2 simulator is used to simulate the results. Here different snapshot shows the result of the research. Following are the results and their brief descriptions.

3.1 Network Setup

For this, two mobile nodes are taken, one is the sender node where the packet is to be forwarded and the other is destination node to which we have to forward the data packets. Different HA (Home Agents) are present in different AS (Autonomous System) and each AS has its BGP router. These routers are responsible for forwarding the data packets.

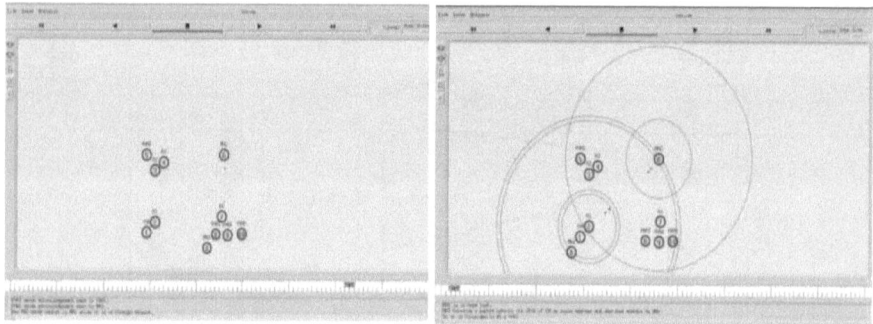

Figure 3: Network setup   **Figure 4:** Simple Mobile IP

3.2. In Mobile IP Scenario

In this result, it shows that how the data packets are forwarding in the Mobile IP scenario. The sender node will send the data packets to the correspondent node through its HA (Home Agent). When the correspondent mobile node is in foreign network then it will get the CoA and this CoA will be announced to all BGP routers and by the help of this ,the sender node will send the packets to CN.

3.3. Registration

In this result, it shows that when the correspondent node moves away from its home network ,it sends the BU (Binding Update) message to its original HA for new HA and finally register with other AS. Now, all the data packets will be send to its new address called as CoA. This is the new address for correspondent mobile node for its further communication.

3.4 Roaming

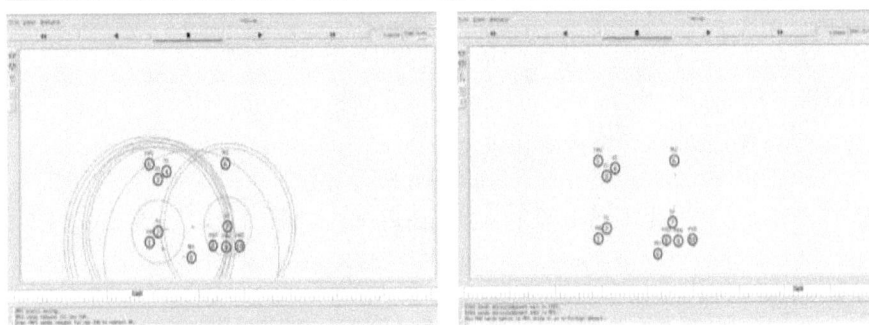

Figure 3: Registering with new AS   **Figure 4:** Roaming Scenario

3.5 Test Cases

Following are the Test cases:

* In the first case, it shows the packet delivery rate. At different time instants, packet delivery rate is being displayed in the following graph.

* In the second case, it shows the efficiency of the system. At different time instants, efficiency of the system is being displayed in the following graph.

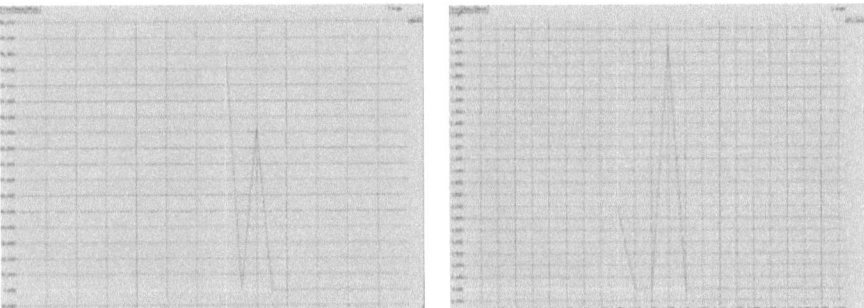

Figure 5: Packet Delivery Ratio   **Figure 6:** Efficiency of the system

4. CONCLUSION

In this paper, by proposing the new architecture it will enhance the mobility of the network and NEMO techniques has aided mobile networks, as it has achieved global reachablity. On top of that when the pre-registration technique is integrated into NEMO, the results were firstly, the packet loss is reduced, which is one of the major factor to increase the efficiency. Secondly, it has reduced the delay, which is another major factor that decides the network efficiency.

5. FUTURE WORK

The architecture can be further enhanced by adding security features that can perform authentication, authorization and accounting during the hand-off. This will enable to keep unauthorized at the bay. This will ensure the network is not compromised and used efficiently.

REFERENCES

[1]   Chun-Hsin Wu et al., "Bi-directional route optimization in mobile IP over wireless LAN", Vehicular Technology Conference, September, 2002.

[2]   G. Huston, "Commentary on inter-domain routing in the internet", RFC 3221, December, 2001.

[3]   K. Lua et al, "A survey and comparison of peer-to-peer overlay network schemes", IEEE Communications Surveys and Tutorials, 2005.

[4]   Marcelo Bagnulo et al., "Scalable Support for Globally Moving Networks", ISWCS, 2006.

[5]   M. Calderon et al., "Design and experimental evaluation of a route optimization solution for NEMO", IEEE JSAC, 2007.

[6]   R. Cuevas, C. Guerrero, A. Cuevas, M. Caldern, C.J. Bernardos, "P2P based architecture for global home agent dynamic discovery in IPmobility", 65th IEEE Vehicular Technology Conference, 2007.

[7]   R. Wakikawa et al.,"Virtual mobility control domain for enhancements of mobility protocols", IEEE INFOCOM, 2006

[8]   T. Clauser et al,"NEMO route optimization problem statement", RFC 4888, October, 2004.

14

E-Commerce Web Application Using Mern Technology

Hemant Kumar Singh[1], Prasoon Mishra[2], Himanshu Baranwal[3]
Department of Computer Science and Engineering, School of Management Sciences, Lucknow
e-mail: 0909prasoon@gmail.com , himanshubaranwal840@gmail.com

ABSTRACT

E-Commerce Shopping Cart is an application that will manage the details of shopping, online payments, customer login, purchases, revenue collection etc. It manages all the information about Shopping, Products and Customer. Through this application user or customer can but their favourite products online by doing quick and easy online payments. Customer can check its cart history whenever required. Admin of this application can check all the details related to customer shopping, customer login or register, sales, payments, revenues etc very easily through admin dashboard. E-Commerce Shopping Cart, as described above, can lead to error free, secure, reliable and fast management system. It can assist the user to concentrate on their other activities rather to concentrate on the record keeping.

Keywords: *React.js, MongoDB, Node.js, Express.js .*

1. INTRODUCTION

We all know that technology has become an essential tool for online marketing these days. If we see all over the world most of the people are showing interest to buy things in online. However, we can see that there are many small shops and grocery stores are selling their things offline. With this type of selling most of us will face bad experience. for instance, in some shops seller has the product to sell in the offer but the buyer may not know about it, or the customer may need the product urgently then he will go to the shop, but the product is out of stock, in that case, he will face bad experience. Moreover, in online shopping customers can select a wide range of products based upon their interests and their price also, one can compare prices also from one store to another by using online shopping . By encountering the all problems and weaknesses of the offline shopping system, creating an E-commerce web-application is necessary for searching and shopping in each shop. These days we have seen so many e-commerce websites are created like Flipkart, Amazon, Myntra one can easily buy their necessary products by using these websites. For creating these types of E-commerce web applications MERN stack will be the best option that can help us for creating the most effective and powerful web applications.

2. PROBLEM DEFINITION

E-commerce provides an easy way to sell products to a large customer base. However, there is a lot of competition among multiple e-commerce sites. When users land on an e-commerce site, they expect to find what they are looking for quickly and easily. Also, users are not sure about the brands or the actual products they want to purchase. They have a very broad idea about what they want to buy. Many customers nowadays search for their products on Google rather than visiting specific e-commerce sites. They believe that Google will take them to the e-commerce sites that have their product. The purpose of any e-commerce website is to help customers narrow down their broad ideas and enable them to finalize the products. They can also buy them easily by just adding to the cart and they can increase or decrease by clicking on the "+" sign and "-" sign. After adding they can check the total amount of the thing which have been added to the cart. A successful payment gateway way enabled so payment can be done by debit card, credit card, and net banking.

3. METHODOLOGY

Customers build up a sense of loyalty to those e-commerce websites that offer them a good user experience, and that transmit to confidence and reliability. There are various factors that influence this: how easy it is to find the product they are looking for, how easy/difficult it is to make the payment, how fast the order was executed. All of these factors determine whether the customer will shop at that website again or not. In general, potential buyers are more and more impatient, which means they do not have much time to find what they are looking for, or to receive a positive first impression. Our Shopping Cart App takes these needs into consideration and as well as others. Our approach is customer-focused and geared towards intuitive web browsing, which combined with a customizable search helps new and returning customers to find the products and services they need, quickly and easily. We also integrate suggestions about similar products to promote other products and increase sales and orders, and thus improve the e-commerce's profitability. When new customers go onto a website that is slow to load, they are quick to get impatient and leave the site. Our e-commerce projects guarantee quick loading, as we use stylesheets (CSS) and files which have been size-optimized. Many users are cautious when shopping online, especially given the regular coverage of online scams on the news. As a result, the web design, usability and content of an ecommerce business needs to offer customers complete peace of mind and convince them of the company's reputation and integrity. We can write or advise you on what to say about customer security on the site. If done well, this will help you increase Work methodology for E-commerce conversions and the site's overall profitability. Our e-commerce module is a trustworthy, secure and user-friendly system that is able to handle high levels of traffic to your website. The payments are made in a very secure way and there is an online tracking facility for the orders. Our specific e-commerce development is user-friendly and can be integrated into any administration system, even with those used by the logistics companies you work with, thus reducing costs and maximizing profits. Web design, usability and natural positioning within search engines. We combine attractive web design for users, simple web browsing, where customers can easily find what they are looking for, with coherence with the company brand image. We have managed a large number of successful e-commerce projects, and in each of them we have adapted our skills in order to meet the client's needs.

4. RESEARCH AND DEVELOPMENT

There are many methodologies to build a web application, in this research, we have used MERN technology for building a web application. MERN: MERN stands for MongoDB, Express.js, React.js, Nodejs. These four technologies help us to construct or to build this web application. MongoDB: It is an open-source cross-platform program. It comes under the NoSQL database classification. It is a document oriented database. It uses JSON format documents with optional Schemas. Data Flexibility available means we can any every data in a separate file .Large data can be distributed into several connected applications .High speed of fetching of data possible because it only depends on indexing. It is a horizontally scalable database so it can handle the data make us easy to distribute to serval machines. NodeJS: NodeJS is a runtime javascript environment that works outside the web page. It is mainly used for server-side applications. NodeJS is open source and it is free of cost. ExpressJS: It is a framework used with NodeJS to increase reusability of codes. ReactJS: It is a frontend library used for making code more compact and divide web pages into codes and reduce redundancy of codes.

5. CONCLUSION

E-Commerce is not just about conducting business transactions via the Internet. Its impact will be far-reaching, and more prominent then we know currently. This is because the revolution in information technology is happening simultaneously with other developments, especially the globalization of the business. The new age global e-commerce is creating entirely new economy and that will tremendously change our lives, will reshape the competition in various industries, and alter the economy globally. As companies are gaining high profits, more and more other companies are developing their websites to increase their profits. Since more businesses are being held online resulting in high economy development and emergence of a more innovative and advanced technology. This web application is easy for them to access and without any effort categories can be created and products can be added by them. It will be very attractive for the customer to see the products by sitting at home or office. It will be very helpful for the small-scale industries without selling to wholesales, large retails mediators they can directly sell to the customer by saving money for both.

ACKNOWLEDGEMENT

It gives us a great sense of pleasure to present the report of the Bachelor of Technology Project undertaking during the final year. We owe special debt of gratitude to the Prof. and (Dean Engg.) (Dr.) Hemant Kumar Singh , (Head of Department) Mr Sunit Mishra, Department of Computer Science & Engineering, School of Management Sciences, Lucknow, affiliated to Dr. A.P.J. Abdul Kalam University, Lucknow for their constant support and guidance throughout the course of our work. Their sincerity, thoroughness and perseverance have been a constant source of inspiration for us. We also do not like to miss the opportunity to acknowledge the contribution of all teaching and non-teaching staff of the Department for their kind assistance and cooperation during the development of our project. We also acknowledge our friends for their contribution, continuous support and encouragement in the completion of the project.

REFERENCES

[1]   Tran, T. T. H. (2022). The development of an e-commerce web application using MERN stack.

[2]   Hoque, S. (2020). Full-Stack React Projects: Learn MERN stack development by building modern web apps using MongoDB, Express, React, and Node. js. Packt Publishing Ltd.

[3]   Sharma, A. K. (2022). BIG BUY (E-COMMERCE) by using MERN.

[4]   Nguyen, H. (2020). End-to-end E-commerce web application, a modern approach using MERN stack.

[5]   Naidu, N. D., Adarsh, P., Reddy, S., Raju, G., Kiran, U. S., Sharma, V., ...& Sharma, V. (2021). E-Commerce web Application by using MERN Technology. International Journal for Modern Trends in Science and Technology, 7, 1-5.

[6]   Suryawanshi, P., Varma, N., Kale, N., &Shrivastav, L. E-COMMERCE WEB APPLICATION USING MERN TECHNOLOGY. Journal homepage: www. ijrpr. com ISSN, 2582, 7421.

[7]   Sharma, N., Kumar, A., Sharma, A., Verma, A., & Srivastava, N. R. E-Commerce Website Using MERN Stack.

[8]   Shukla, S. K., Dubey, S., Rastogi, T., & Srivastava, N. (2022). Application using MERN Stack.

[9]   Mai, N. (2020). E-commerce Application using MERN stack.

15

Green Synthesis of Titania and Titania-Silver Nanoparticle from Novel Plant Extract of Origanum Majorana

Aakash Singh[1*], Ved Kumar[2], Anod Kumar Singh[3], Sudhaker Dixit[4]

[1*]Department of Applied Sciences, IET, Dr. Shakuntla Mishra National Rehabilitation University, Lucknow, Uttar Pradesh, India; e-mail : aakashsinh.ucst@gmail.com

[2-4]Department of Humanities & Applied Sciences, School of Management Sciences, Lucknow

GRAPHICALABSTRACT

TiO_2 Nanomaterials are synthesized by facile, novel, greener, sustainable, fast, and eco-friendly techniques are now-a-days getting wide attention in all the branches of science. Metal oxides nanoparticles are extensively applicable to perform distinctive physiochemical properties in various types of biological applications. Among metal oxide TiO2 is best knownforextensive approaches in nanotechnology, biotechnology and medicine. Our approach is for plant mediated green synthesis of TiO_2 and Ag-TiO_2 nanoparticle by aqueous leaf extract of *Origanummajorana* During the study different result was observed and they were examined by the UV-visible spectrum as the basic characterization and further formation of Titanium oxide NPs were confirmed by FTIR, SEM, and XRD. The method was eco friendly and less hazardous as compared with other conventional methods,

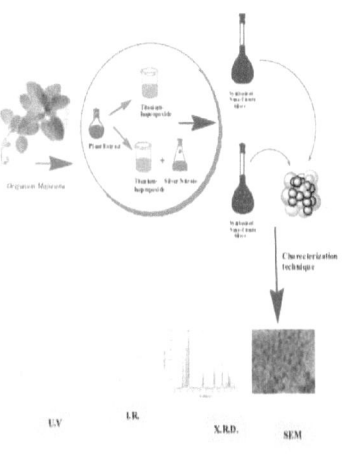

Key words :*GreenSynthesis, Origanummajorana, Titanium Oxide nano particles.*

1. INTRODUCTION

Nanotechnology was witnessed to achieve attention in various fields of science and technology right from medicinal to biological and pharmaceutical to industrial since last two decades. Various practices were made to achieve the goal of nano-sciences by using various materials either independently or being doped, but due to its unique property titania powder getting more and more attention in form of nanotitaniumdioxide and silver supported titanium nano particles. They posses wide range of applications like medicinal (application in drugs, dyes, pigments, paints, catalyst and photo catalyst) morphological properties (due to shape, size and crystalline nature) [1-5] and one of the significant property is numbers of preparation methods (like sol-gel method,[6-9]solvothermal method,[10,11] hydrothermal method,[12,13] mechanical alloying, milling,[14,15]mechano-chemical method,[16,17] chemical, plant based reduction method and RF thermalplasma method).[18]

Recently TiO_2 nanoparticles were synthesized using variety of biosynthetic processes and naturally occurring sources like *Lactobacillus sp.* And Schharomycescerevisae (having antibacterial and antifungal activity), [19] *Nyctanthesarbortristis*extract, [19]*catharanthusroseus*extract, [20]*Eclipta prostate* [21]*and Aspergillus flavus,*[22] *Annona squamosa peal* extract,[23] *Azadirachtaindica* [24]and so on.

Ag-TiO_2 due to its nano sized dimensional occurrence have gained wide significance in field of chemical stability, non-toxicity, skin care wareses, skin protection against UV, cosmetics industries, gas sensors, electro-chromic devices, bone tissue engineering, sunscreen, photo catalyst, dielectric, plastic, paper, inc, food colorant and toothpaste etc. [25-29]

Nano Ag-TiO_2/ TiO_2 particles have their characteristic properties like antibacterial, antimicrobial, antifungal because the functionalized nanoparticles shows improved inhibition against bacteria, fungus and other microbes.[30]

Origanummajorana is known as *Marua* in India. The sweet *majorana* is pleasant smelling herb which belongs to the mint family having small leaf and white flowers. *Origanummajorana* is medicinally valuable in flatulence, nausea, abdominal bloating and minor neural problems. The plant also has some therapeutic values such as used as a tonic and as effective stimulants, control excessive secretion as well as bleeding, loss of appetite, intestinal spasm, very effective sedative, relieving indigestion caused due to neural issues, palpitation and insomnia, calmative uses, stimulates lactation and perspiration. In Ayurved it is used to tooth ache, sooth joints and muscular pain. [31-33]

The main chemical constituents of *Origanummajorana* found were: *cis*-sabinene hydrate (30.2%), terpinen-4-ol (28.8%), γ-terpinene (7.2%), á-terpineol (6.9%), *trans*-sabinene hydrate (4.4%), linalyl acetate (3.8%), and α-terpinene (3.6%).

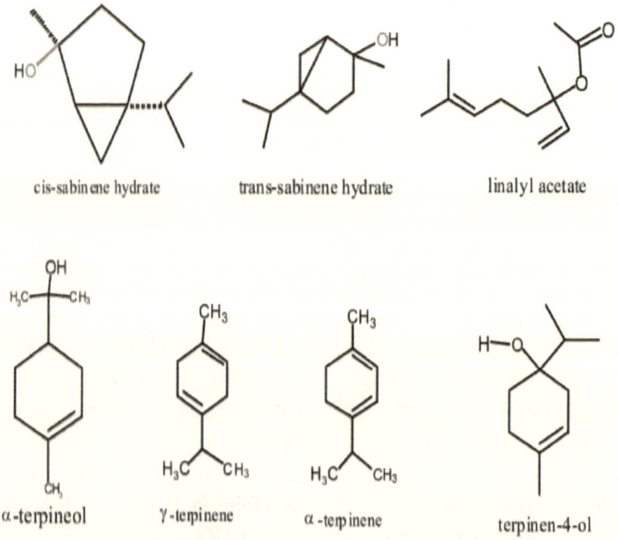

cis-sabinene hydrate   trans-sabinene hydrate   linalyl acetate

α-terpineol   γ-terpinene   α-terpinene   terpinen-4-ol

Scheme 1: chemical constituents of Origanummajorana

The plant also has very good nutritional composition as tannins, manganese, zinc and flavonoids, and essential oils which are rich in thymol, sabinene, terpenes, linalool and cavacrol.

2. MATERIAL AND METHODS

All the chemical are of analytical grade titanium isopropoxide (s.d. fine chem) Ethanol (Merck) Oleic acid (Merck), AgNO₃ (Spectrochem), UV Spectrophotometer double beam Shimadzu model no 1800, FTIR Perkin Elmer Spectrum Version 10.4.00 (4000-400 cm⁻¹ KBr) SEM Nova Nano, XRD (PanalyticalXpert pro)

3. PREPARATION OF PLANT EXTRACT SOLUTION

Fresh leaves of *OriganumMajorana* were collected from the local area of Jaipur District, Rajasthan, India. The collected leaves were washed carefully under running water to remove the dust particle. 5 gm of fresh leaves were cuted very fine and boil in 100 ml of water for about 2-3 hrs at 60-70^{0C}. The leaf extract when changes to dark brown color then allowed to stand settle down for overnight and filter the extract and store the extract solution bellow 20^{0C} before use.

4. SYNTHESIS OF TITANIUM OXIDE NANOPARTICLE

0.5 M of titaniumisopropoxide added to 50ml of plant extract solution. Fix the pH of solution at one by adding HNO₃. The acidic solution of titaniumisopropoxide and leaf extract of *OriganumMajorana* heated at 60^{0C} for 2 hrs under vigorous stirring. After two hrs continuous string allowed the solution to settle down for 6 hrs. Remove the watery part if any and centrifuge the rest of solution at 15000 rpm for 10 min to obtain the nano TiO2 particles. The collected metal nanoparticles are the washed repeatedly 2-3 times with ethanol and water and then centrifuge again at 10000 rpm, finally separated nanoparticles are then dried in hot air oven at 80^{0C} and then calcinate to 700^{0C} for 4 hrs in MuffalFurnance grinned the metal nanoparticles using Motel-Pistal to get the finest particle size suitable for further Characterization technique.

5. SYNTHESIS OF NANO TITANIA-SILVER

1.5 gm of nano titanium oxide particle mixed with1.5 gm of 1×10^{-2} M oleic acid solution in methanol under string 500 rpm at room temperature until the complete evaporation occurs, 0.1M AgNO₃ solution mixed with the dried sample of oleic acid TiO₂ and string again for 36 hrs at 500 rpm under dark at room temperature, color change was observed from yellowish white to dark gray color, the product was then dried at100^{0C} in hot air oven, washed repeatedly 2-3 times with large amount of ethanol and water. Nanoparticle sample were calcinated at 700^{0C} in muffle Furnace and then grind the sample using Motel Pistal to crush the particle into finest particle size for better characterization.

6. RESULTS AND DISCUSSION

6.1 Plausible mechanism for formation of TiO2 nanoparticles

Chemical composition of *OriganumMajorana*favours the formation of nano TiO₂ it contains appreciable amount of *cis* and *trans* -sabinene hydrate, terpinen-4-ol, α-terpineoland linalyl acetate. The hydroxyl group present as a functional group in the bio-organic phytochemical constituents under plant extracts to dehydrate titanium isopropoxide to Tio₂ nanoparticle by refluxing it with plant extract solution for 30-40 min at 60-70 °C. Terpinen-4-ol, α-terpineol and sabinene hydrateservs as a catalyst in thermal decomposition reaction due to the breakdown of titanium isopropoxide $[Ti(OCH(CH_3)_2)]_4$ into titanium hydroxide $Ti(OH)_4$ and cyclopropanol$(CH_2)_2CHOH$ further condensation reaction removes water molecule and form titanium oxide nanoparticles which on further calcinated and studied (Figure 1).

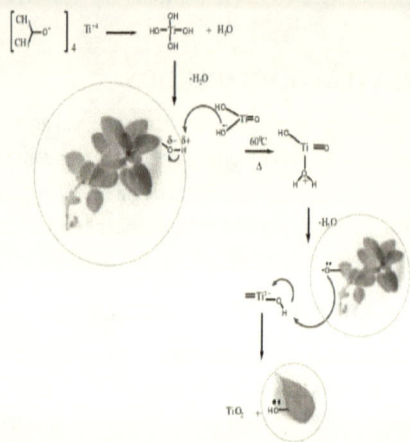

Figure 1: Plausible route for the dehydration of titanium isopropoxide to TiO_2 nanoparticles

The toxicity [34] and applicability [35] of TiO_2 and doped TiO_2 nanoparticles has received considerable attention because of the fascinating ability of TiO_2 nanoparticles to produce reactive oxygen species in separately as well as attached with other metals. As a part of our continuing effort for the development of efficient green methodology in the field of biosynthesis of nanoparticles we have also report herein the doping of silver with TiO_2 nanoparticles and form titania silver nanoparticles.

TiO_2NPs formation reaction can be predicted as below: [36]

Hydrolysis:
$$Ti(OCH(CH_3)_2)_4 + 4H_2O \longrightarrow Ti(OH)_4 + 4(CH_2)_2CHOH.... (1)$$

Condensation:
$$Ti(OH)_4 \longrightarrow TiO_2 \cdot xH_2O + (2-x)H_2O ....(2)$$

Crystallisation via calcination:
$$TiO_2 \cdot xH_2O \longrightarrow TiO_2 ....(3)$$

where x = number of water molecules.

The OH group favors the hydrolysis of titanium isopropoxide followed by condensation which formed an unstable intermediate $Ti(OH)_4$ and precipitation takes place. The white precipitate occurs due to formation of $TiO_2.x\ H_2O$ as shown in equation (2). This precipitate goes through calcinations in muffle furnace at temp range 200-700 °C and shows various size dependent changes.

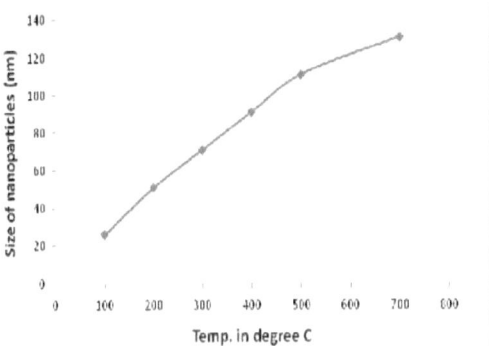

Figure 2: Temperature Size Graph

The crystalline size of TiO_2 nanoparticles depends upon the calcinating temperature as the temperature increases size of nanoparticles also increases. Thermally promoted crystalline growth can observed when temperature ranges from 200-500 °C and also about 700 °C the particle size become very large exceeded 150 nm (figure 2). Nucleation occurs in growth of TiO_2 nanoparticles at higher temperature and shows aggregation pattern in SEM analysis (figure 5 and 6)

6.1 Ultraviolet Spectroscopy

UV–Vis DRS spectra were obtained to investigate the light reflectance and absorption characteristics of the electronic states of TiO2 nanoparticles. Figure. 4 shows the UV–Vis DRS of TiO2 nanoparticles with the maximum reflectance and absorbance around at 317 nm and 357 nm (Figure. 3)Peaks obtained at green range absorption wavelength357.50, 7.00, 281.00 [37,38]

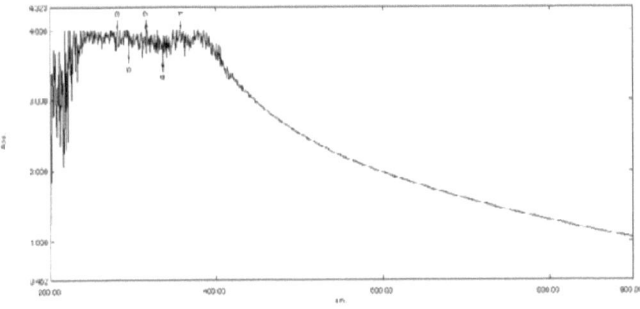

Figure 3: UV visible spectroscopy of TiO_2 Nanoparticles

Fourier transformation infrared spectroscopy (FTIR) of TiO2 nano and Ag-TiO2 nano sample were recorded from Perkin Elmer spectrum version 10.4.00 (4000-400 cm^{-1}) using KBr pallets. In the graph line at T1 (Figure 4a) various peaks at 3435.91cm-1 (H-O-H str), 2921.41 cm^{-1}(C-H str), 2863.30 cm^{-1} (C-H str), 2078.15 cm^{-1}(C=Ostr) confirms the different capping agents present in plant extract which get merged with the surface of nanoparticles. Whereas peaks at 651 cm^{-1} and 496 cm^{-1} confirms the presence of TiO$_2$.

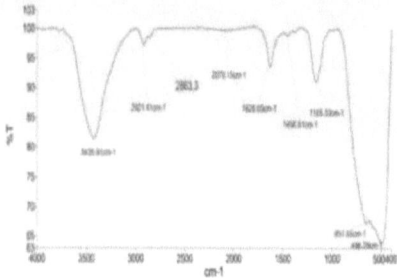

| **Figure 4a:** I.R spectrum of TiO2 NPs | **Figure 4b:** I.R spectrum of Ag-TiO2 NPs |

In I.R spectrum of Ag-TiO$_2$ nanoparticle the graph line (Figure 4b) confirms the various peaks of capping agents attached with the surface of nanoparticles at 3437 cm^{-1} (H-O-H str), 2919 cm^{-1} (C-H str), 2315 cm^{-1}(Ca"C and Ca"N), 1624 cm^{-1}(silver) along with 1554 and 1438 cm^{-1}. Peaks at 605.71 and 522.27 cm^{-1} confirms anatase phase TiO$_2$[39]

6.2 Scanning electron microscopy and EDX

(a)   (b)

Figure 5 : (a,b) - SEM images of TiO2 nanoparticles

Sem analysis shows that TiO$_2$nanoparticles formed clusters of aggregates, at scale bar of, 300 nm and 1µm they appeared as accumulated shape with particle size less than 100 nm. (Figure: 5a,b) The silver supported nano particle Ag-TiO$_2$are at scale bar 500 nm confirms the nano range of particle having size 51 nm. They are dumble in shape. The particles are heterogeneous in nature but nano range clearly appear in the SEM image. (Figure: 6a,b) [40]

(a)   (b)

Figure 6: (c,d)- SEM images of Ag-TiO$_2$ nanoparticles

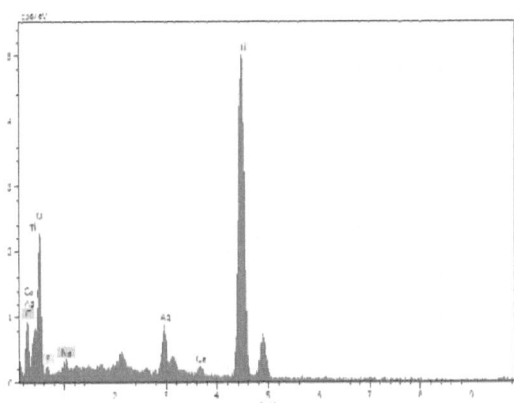

Figure 7: EDX pattern of Ag-TiO$_2$ nanoparticles

Electron dispersive x-ray spectroscopy shows sharp peak of Ti and Ag which confirms the formation of metal nanoparticles and their attachment with silver metal. Rest of the element like Ca, C, Na, F, detected by EDX characterization was due to plant related alkaloid attached to the surface of nanoparticles.

6.3 X-ray Diffraction Spectroscopy

The XRD analysis of synthesized nano TiO$_2$ from plant extract confirms that they are in anatase phase having colorless tetragonal crystal geometry. Nine distinct peak of 2Ÿ = 25.60, 38.80, 48.25, 54.07, 55.28, 62.87, 69.07, 70.49, 76.26 the values were compared with the standard (JCPDS fill no-21-1272). (Figure: 8a)

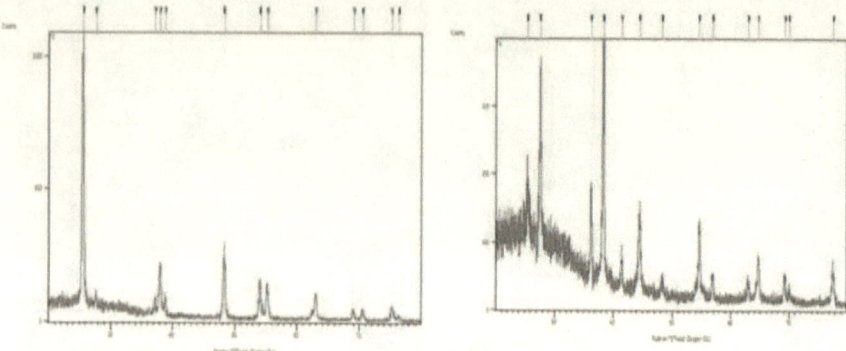

Figure 8a: XRD plot of TiO$_2$ nanoparticles **Figure 8b:** XRD plot of Ag-TiO$_2$ nanoparticles

In Ag-TiO$_2$ nanoparticle they are in silver-3C having light gray color metallic cubic crystal geometry. Four distinct peaks of 2Ÿ = 38.32, 44.44, 64.58, 77.58 the values were compared with the standard (JCPDS fill no-04-0783). (Figure: 8b) The sharpness of peaks and the absence of unidentified peaks confirmed the crystalline and high purity of nanoparticles prepared. The average crystalline size of TiO2 nanoparticles was calculated using Debye-Scherrer's equation.

$$D = K\lambda / \beta\cos\theta \ldots\ldots\ldots\ldots\ldots\ldots\ldots\ldots\ldots\ldots\ldots.equation(1)$$

where D - average crystal size
λ - wavelength of the X-ray radiation
K - dimensionless shape factor (Cu Kα took as 0.9)
β - line width at half-maximum intensity (FWHM)
θ - Bragg's angle.
The average crystalline size (D) using this formula was ± 20 nm.

7. BIOLOGICAL ACTIVITY

7.1 Antibacterial Activity

Well diffusion method was used to examine the antibacterial activity of formed nanoparticles against the model test strain of gram negative *E. coli* (MTCC 1721) and gram positive *S.aureus* (MTCC 3160). The sterilizing temperature, pressure and time for all the media, glassware and reagent are 125^{0C}, 15 PSI and 35 min respectively. The utilized bacterial concentration 1.8×10^8 CFU/ml was tested against 20 ml of three samples-*Origanummajorana*Plant extract, TiO$_2$Nps, and Ag-TiO$_2$Npsat three different concentration 12.5 mg/L, 25 mg/L and 50 mg/L in three different steps. The plates were incubated for 24 hrs at 35-37^{0}C temp.

The final result of antimicrobial activities was observed by measuring the growth of zone of inhibition diameter in mm and MIC in µgm/ml for plant extract, plant extract + TiO_2Npsand plant extract + Ag-TiO_2Npsas given in (Table 1). Ag-TiO_2Npshave maximum growth of zone of inhibition at various concentrationsi.e., 12.5, 25, 50 mg/disc. The zone of inhibition was measured as 13.53mm, 18.00mm and 20 mm respectively for *S. aureus* and comparatively less growth for*E. coli* was found to be 11.50 mm, 16.00 mm and 18.00 mm respectively (Fig 9 a, b and c). The result of Antibacterial activity revealed that of Ag dopedtitanium oxide nanoparticles observed to be more active than plant extract and TiO_2 and was more potent against *S. aureus* with 50mg/l value.

Table-1: Observation of antibacterial activity

S. N	Name of Bacteria	Test Sample	Concentration	Zone of inhibition (mm)
1	*E. coli* (MTCC 1721)	*Origanummajorana*extract	12.5 mg/L	4
			25 mg/L	4
			50 mg/L	6
2	*E. coli* (MTCC 1721)	TiO_2Nps	12.5 mgLl	7.5
			25 mg/L	6
			50 mg/L	11
3	*E. coli* (MTCC 1721)	Ag-TiO_2Nps	12.5 mg/L	11.5
			25 mg/L	16
			50 mg/L	18
4	*S. aureus* (MTCC 3160)	*Origanummajorana*extract	12.5 mg/L	5.5
			25 mg/L	6
			50 mg/L	7.5
5	*S. aureus* (MTCC 3160)	TiO_2Nps	12.5 mg/L	9.5
			25 mg/L	8
			50 mg/L	12.5
6	*S. aureus* (MTCC 3160)	Ag-TiO_2Nps	12.5 mg/L	13.5
			25 mg/L	18
			50 mg/L.	20

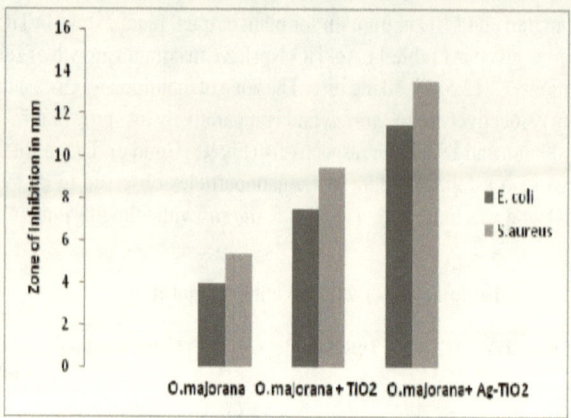

Figure 9a : Graph at concentration 12.5 mg/L

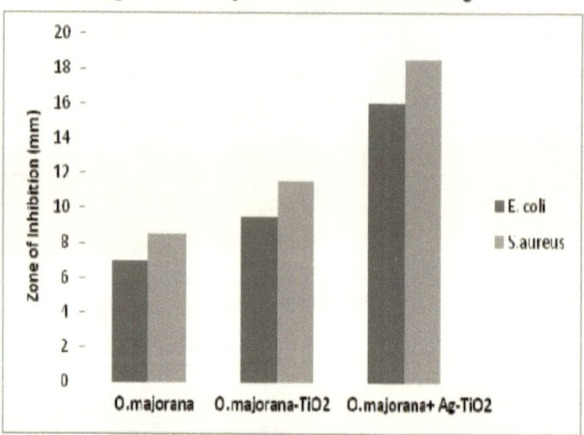

Figure 9b : Graph at concentration 25 mg/L

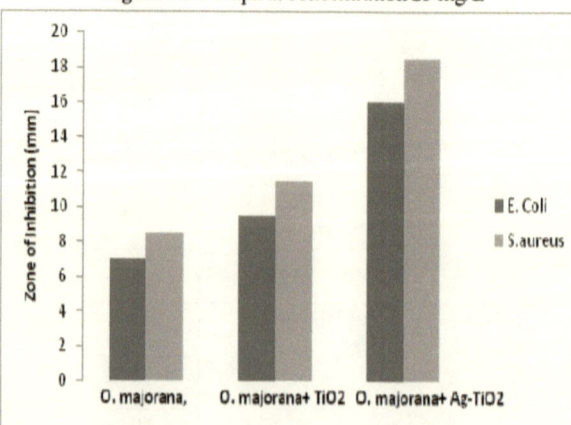

Figure 9c : Graph at concentration 50 mg/L

7.2 Antifungal Activity

The Antifungal activity of *Origanummajorana* plant extract, *O.majorana* +TiO$_2$Nps and *O.majorana* + Ag-TiO$_2$Nps to be tested against *Aspergillus niger* fungus by well diffusion method on the surface of PDA (Potato Dextrose Agar)media incubated with 1×10^5 CFU/ml of spore suspension of *Aspergillus niger* fungi.

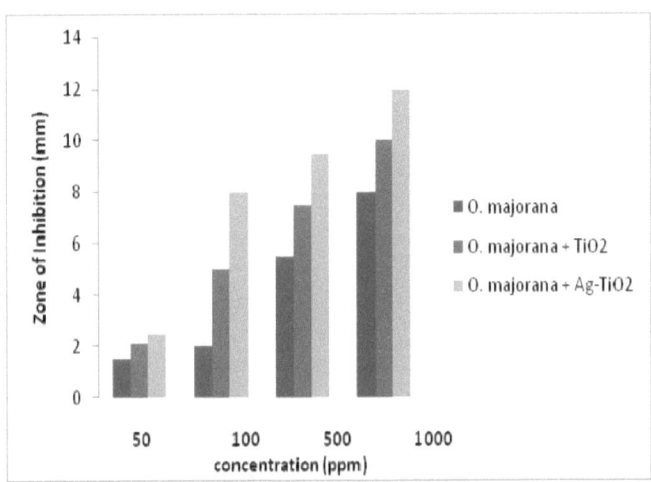

Figure 10: Comparison of Antifungal activity of Aspergillus *niger* at conc. (50ppm, 100ppm, 500 ppm, 1000 ppm).

The potato dextrose broth (PDB) was prepared and placed in shaking incubator for 48 hrs and then *Aspergillus niger* fungus were suspended into it. Potato Dextrose Agar (PDA) is allowed to set and were uniformly seeded with *Aspergillus niger* fungus taken from suspension.

Small wells having 6mm diameter impregnated with the solution of *Origanummajorana* plant extract, TiO$_2$Nps and Ag-TiO$_2$Nps were poured into the plates of culture medium at different concentration of 50, 100, 500 and 1000 ppm. Plates were immediately transferred to incubator for 48 hrs at 25-28^{0}C temp. After 48 hrs of incubation, degree of sensitivity is determined by measuring the zone of inhibition in mm. (Table 2). The increasing ppm concentration is directly proportional to activity against *Aspergillus niger*. The best activity was shown by Ag-TiO$_2$Nps of *Origanummajorana* at 1000 ppm concentration.(Figureure 10).

Table-2: Observation of antifungal activity

S. N	Name of Fungi	Concentration (ppm)	Test Sample	Zone of inhibition (mm)
1	*Aspergillus niger*	50	*O.Majorana extract*	1.5
			TiO_2Nps	2.1
			$Ag-TiO_2Nps$	2.5
2	*Aspergillus niger*	100	*O.majorana*extract	2
			TiO_2Nps	5
			$Ag-TiO_2Nps$	8
3	*Aspergillus niger*	500	*O.Majorana*extract	5.5
			TiO_2Nps	7.5
			$Ag-TiO_2Nps$	9.5
4	*Aspergillus niger*	1000	*O.Majorana*extract	8
			TiO_2Nps	10
			$Ag-TiO_2Nps$	12

8. CONCLUSIONS

In summary, we have developed convenient and green procedure for the synthesis of TiO_2 and silver supported TiO_2 from novel plant *Origanummajorana* for the first time. The employed plant plays important role as capping agent on the surface of metal nanoparticle of $Ag-TiO_2$ and bioactivity against microbial and fungal strains recorder up to a good significant results.

ACKNOWLEDGEMENTS

A.S. thanks Research laboratory Institute of Engineering and Technology, Dr. Shakuntla Mishra National Rehabilitation UniversityLucknow for experimental performance.Author also thankful to the USIC, University of Rajasthan Jaipur and MRC, MNIT, Jaipur for the spectral analysis.

REFERENCES

[1]   C.J. Barbe, F. Arendse, P. Comte, M. Jirousek, M. Gr̈atzel, Nanocrystalline titanium oxide electrodes for photovoltaic applications, J. Am. Ceram. Soc. 80 (1997) 3157.

[2]   R. Monticone, A.V. Tufeu, E. Kanaev, C. Scolan, Sanchez, Quantum size effect in TiO2 nanoparticles: does it exist, Appl. Surf. Sci. 162–163 (2000) 565–570.

[3]   S. Boujday, F.Wunsch, P. Portes, J.-F. Bocquet, C.C. Justin, Photocatalytic and electronic properties of TiO2 powders elaborated by sol–gel route and supercritical drying, Solar Energy Mater. Solar Cells 83 (2004) 421–433.

[4]   O. Carp, C.L. Huisman, A. Reller, Photoinduced reactivity of titanium dioxide, Prog. Solid State Chem. 32 (2004) 133–177.

[5]   A.M. Ruiz, G. Sakai, A. Cornet, K. Shimanoe, J.R. Morante, N. Yamazoe, Microstructure control of thermally stable TiO2 obtained by hydrothermal process for gas sensors, Sens. Actuators B: Chem. 103 (2004) 312–317.

[6]   T. Trung, W.-J. Cho, C.-S. Ha, Preparation of TiO2 nanoparticles in glycerol-containing solutions, Mater. Lett. 57 (2003) 2746–2750.

[7]   T. Sugimoto, X. Zhou, A. Muramatsu, Synthesis of uniform anatase TiO2 nanoparticles by gel–sol method. 1: Solution chemistry of Ti(OH)n ($4"n$)+ complexes, J. Colloid Interface Sci. 252 (2002) 339–346.

[8]   T. Sugimoto, X. Zhou, A. Muramatsu, Synthesis of uniform anatase TiO2 nanoparticles by gel–sol method. 2: Adsorption of OH" ions to Ti(OH)4 gel and TiO2 particles, J. Colloid Interface Sci. 252 (2002) 347–353.

[9]   P. Arnal, R.J.P. Corriu, D. Leclercq, P.H. Mutin, A. Vioux, A solution chemistry study of nonhydrolytic sol–gel routes to titania, Chem. Mater. 9 (1997) 694–698.

[10]   C.-S. Kim, B.K. Moon, J.-H. Park, S.T. Chung, S.-M. Son, Synthesis of nanocrystalline TiO2 in toluene by a solvothermal route, J. Cryst. Growth 254 (2003) 405–410.

[11]   C.-S. Kim, B.K. Moon, J.-H. Park, B.-C. Choi, H.-J. Seo, Solvothermal synthesis of nanocrystalline TiO2 in toluene with surfactant, J. Cryst. Growth 257 (2003) 309–315.

[12]   J.-N. Nian, H. Teng, Hydrothermal synthesis of single-crystalline anatase TiO2 nanorods with nanotubes as the precursor, J. Phys. Chem. B 110 (2006) 4193–4198.

[13]   Y.V. Kolen'ko, B.R. Churagulov, M. Kunst, L. Mazerolles, C. Colbeau-Justin, Photocatalytic properties of titania powders prepared by hydrothermal method, Appl. Catal. B: Environ. 54 (2004) 51–58.

[14]   D.H. Kim, H.S. Hong, S.J. Kim, J.S. Song, K.S. Lee, Photocatalytic behaviors and structural characterization of nanocrystalline Fe-doped TiO2 synthesized by mechanical alloying, J. Alloys Compd. 375 (2004) 259–264.

[15]   P. Xiaoyan, M. Xueming, Study on the milling-induced transformation in TiO2 powder with different grain sizes, Mater. Lett. 58 (2004) 513–515.

[16]   J.L. Guimaraes, M. Abbate, S.B. Betim, M.C.M. Alves, Preparation and characterization of TiO2 and V2O5 nanoparticles produced by ball-milling, J. Alloys Compd. 352 (2003) 16–20.

[17]   M. Kamei, T. Mitsuhashi, Hydrophobic drawings on hydrophilic surfaces of single crystalline titanium dioxide: surface wettability control by mechanochemical treatment, Surf. Sci. 463 (2000) L609–L612.

[18]   S.-M. Oh, T. Ishigaki, Preparation of pure rutile and anatase TiO2 nanopowders using RF thermal plasma, Thin Solid Films 457 (2004) 186–191.

[19]   M. Sundrarajan, S. Gowri, Chalcogenide Letters (2011) 447–451.

[20]   K. Velayutham, A.A. Rahuman, G. Rajakumar, T. Santhoshkumar, S. Marimuthu, C. Jayaseelan, A. Bagavan, A.V. Kirthi, C. Kamaraj, A.A. Zahir, G. Elango, Parasitol. Res. (2011), http://dx.doi.org/10.1007/s00436-011-2676-x.

[21]   A.V. Kirthi, A.A. Rahuman, G. Rajakumar, S. Marimuthu, T. Santhoshkumar, C. Jayaseelan, G. Elango, A.A. Zahir, C. Kamaraj, A. Bagavan, Mater. Lett. 65 (2011) 2745–2747.

[22]   G. Rajakumar, A.A. Rahuman, S.M. Roopan, V.G. Khanna, G. Elango, C. Kamaraj, A.A. Zahir, K. Velayutham, Spectrochim. Acta A91 (2012) 23–29.

[23]   Roopan, SM; Bharathi, A.; Prabhakarn, A.; Rahuman, AA; Velayutham, K. and Rajakumar, G. (2012). Efficient phyto-synthesis and structural characterization of rutile TiO2 nanoparticles using Annona squamosapeel extract. SpectrochimActa. A Mol. Biomol. Spectrosc. 98: 86-90.

[24] Siegel, RW.; Hu, E. and Roco, MC. (1999). Nanostructure Science and Technology: R & D Status and Trends in Nanoparticles, Nanostructured Materials, and Nanodevices. Kluwer Academic Publishers, Boston, pp. 336.

[25] Hoffmann MR, Martin ST, Choi WY, Bahnemann DW. Environmental applications of semiconductor photocatalysis. Chem Rev 1995; 95: 69-96.

[26] Fujishima A, Rao TN, Truk DA. Titanium dioxide photocatalysis. J PhotochemPhotobiol C: Photochem 2000; 1: 1-21.

[27] Gelis C, Girard S, Mavon A, Delverdier M, Pailous N, Vicendo P. Assessment of the skin photo protective capacities of an organomineral broad spectrum sunblock on two ex vivo skin models. PhotodermatolPhotoimmunolPhotomed 2003; 19: 242-253.

[28] Trouiller B, Reliene R, Westbrook A, Solaimani P, Schiestl RH. Titanium dioxide nanoparticles induce DNA damage and genetic instability in vivo in mice. Cancer Res 2009; 69: 8784-8789.

[29] Gong P, Li H, He X, Wang K, Hu J, Zhang S, et al. Preparation and antibacterial activity of Fe3O4@ Ag nanoparticles. Nanotechnology 2007; 18: 604-611.

[30] Allahverdiyev AM, Abamor ES, Bagirova M, Rafailovich M. Antimicrobial effects of TiO(2) and Ag(2)O nanoparticles against drug-resistant bacteria and leishmania parasites. FutMicrobiol 2011; 8: 933-940.

[31] Roby MHH, Sarhan MA, Selima KAH, et al. Evaluation of antioxidant activity, total phenols and phenolic compounds in thyme (Thymus vulgaris L.), sage (Salvia officinalisL.), and marjoram (OriganummajoranaL.) extracts. Ind Crop Prod. 2013;43(2):827-831

[32] Zaidi SF, Yamada K, Kadowaki M, et al. Bactericidal activity of medicinal plants, employed for the treatment of gastrointestinal ailments, against Helicobacter pylori. J Ethnopharmacol. 2009;121(2):286-291

[33] el-Ashmawy IM, el-Nahas AF, Salama OM. Protective Effect of Volatile Oil, Alcoholic and Aqueous Extracts of Origanummajoranaon Lead Acetate Toxicity in Mice. Basic ClinPharmacolToxicol. 2005;97(4): 238-243.

[34] Long TC, Saleh N, Tilton RD, Lowry GV, Veronesi B. Titanium dioxide (P25) produces reactive oxygen species in immortalized brain microglia (BV2): implications for nanoparticle neurotoxicity. Environ SciTechnol 2006; 40: 4346-4352.

[35] Badireddy AR, Hotze EM, Chellam S, Alvarez P, Wiesner MR. Inactivation of bacteriophages via photosensitization of fullerol nanoparticles. Environ SciTechnol 2007; 41: 6627-6632.

[36] S. Mahshid, Synthesis of TiO2 nanoparticles by hydrolysis and peptization of titanium isopropoxide solution, J.Mater. Process. Technol., 2007, 189, 296-300.

[37] Shu-Ya Du1,2 and Zhi-Yuan Li1,* optics letters / Vol. 35,issue No. 20 /p.g 3402/ 2010

[38] Swayamprava Dalai, SunandanPakrashi, Suresh Kumar R. S., N. Chandrasekaran, Amitava Mukherjee*/ Material for Toxicology Research /Royal Society of Chemistry 2012.

[39] Silverstein RM, Webster FX, Kiemle D. Spectrometric identification of organic compounds. 7th edition. John Wiley&Sons; 2005, p. 502.

[40] D.Rachel Evangelene Tulip, Aishwarya, K.K. Surya, K. Krishna Devi, R. Kousalya, Biosynthesis of silver nanoparticles using Morindacitrifolia L. as capping and reducing agents, IJETT 3 (2012) 24-34.

www.ingramcontent.com/pod-product-compliance
Lightning Source LLC
Chambersburg PA
CBHW020316290526
45785CB00007B/2817